Skillmaths 1

Skillmaths 1

STAN A. ALTHANS, B.Sc.

formerly Senior Lecturer in Mathematics,
Newland Park College of Education, Buckinghamshire
now teacher of mathematics, St. Margaret's School,
Bushey, Herts

and

MARGERY DENT, B.Sc.

formerly Senior Lecturer in Mathematics,
Newland Park College of Education, Buckinghamshire

Macmillan Education
London and Basingstoke

First published 1983
Published by
MACMILLAN EDUCATION LIMITED
Houndmills Basingstoke Hampshire RG21 2XS
and London
Associated companies throughout the world

Typeset in Hong Kong by Asco Trade Typesetting Ltd.
Printed in Hong Kong

British Library Cataloguing in Publication Data
Althans, Stanley
Skillmaths 1
1. Mathematics—1961–
I. Title   II. Dent, Margery
510   QA39.2
ISBN 0-333-32222-3

Contents

Odd and even

A Take some centimetre squared paper.
Copy these ODD number shapes and then cut them out.
Count the squares.

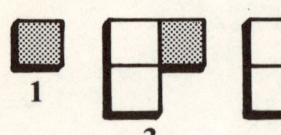 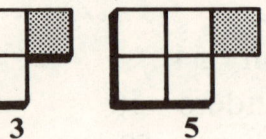 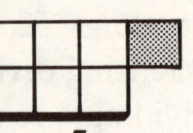

Continue these shapes up to 15.

Write the set of ODD numbers   {1, 3, 5, 7 . . . 15}

B Copy the EVEN number shapes below on to your squared paper.
Cut them out.   Count the squares.

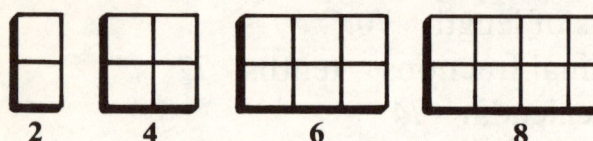

Continue these shapes up to 16.

Write the set of EVEN numbers   {2, 4, 6, 8 . . . 16}

C

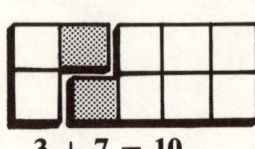

3 + 7 = 10

Two ODD numbers added together make an EVEN number.

Draw sketches on your squared paper, or use your cut out shapes to check this.

Try:

1. 1 + 3 = ☐   **2.** 5 + 3 = ☐   **3.** 1 + 9 = ☐   **4.** 5 + 9 = ☐
5. 7 + 5 = ☐   **6.** 3 + 11 = ☐   **7.** 9 + 7 = ☐   **8.** 7 + 1 = ☐
9. 11 + 5 = ☐   **10.** 9 + 3 = ☐   **11.** 3 + 3 = ☐   **12.** 7 + 7 = ☐

D

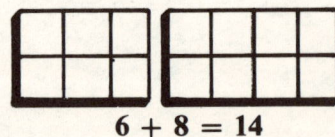

6 + 8 = 14

Two EVEN numbers added together also make an EVEN number.
Use your shapes to check this.

Try:

1. 4 + 6 = ☐   **2.** 2 + 4 = ☐   **3.** 6 + 10 = ☐   **4.** 6 + 2 = ☐
5. 8 + 4 = ☐   **6.** 8 + 2 = ☐   **7.** 6 + 12 = ☐   **8.** 8 + 6 = ☐
9. 4 + 10 = ☐   **10.** 12 + 4 = ☐   **11.** 14 + 2 = ☐   **12.** 8 + 8 = ☐

E Try adding an ODD number to an EVEN number.

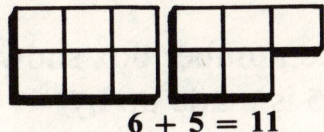

6 + 5 = 11

Use your shapes to try:

1. 4 + 3 = □   **2.** 6 + 7 = □   **3.** 2 + 5 = □   **4.** 2 + 9 = □
5. 4 + 11 = □   **6.** 8 + 3 = □   **7.** 8 + 7 = □   **8.** 6 + 9 = □
9. 8 + 5 = □   **10.** 12 + 7 = □   **11.** 6 + 13 = □   **12.** 10 + 9 = □

What kind of number is the answer in each sum?

F Now add an EVEN number to an ODD number.

1. 3 + 2 = □   **2.** 5 + 4 = □   **3.** 7 + 4 = □   **4.** 5 + 6 = □
5. 9 + 4 = □   **6.** 7 + 6 = □   **7.** 5 + 8 = □   **8.** 9 + 6 = □
9. 7 + 8 = □   **10.** 9 + 10 = □   **11.** 11 + 2 = □   **12.** 7 + 10 = □

G Look back at your results.
Copy and complete this table.

EVEN + EVEN = _____
 ODD + EVEN = _____
 ODD + ODD = _____
EVEN + ODD = _____

H Large ODD numbers end with an ODD number   (1, 3, 5, 7, 9)
Large EVEN numbers end with an EVEN number   (0, 2, 4, 6, 8)
1 7⑤ is ODD   2 9⑥ is EVEN

Sort these into sets of EVEN and ODD numbers.

42	53	61	47	78	80	31	29
126	259	531	324	256	728	196	353
729	840	333	245	625	900	374	100

EVEN = {42, ...}   ODD = {53, ...}

What must you add to an ODD number to make it EVEN?

Hundreds

A

hundreds	tens	ones
• • •	• •	• • • • •
3	**2**	**5**

The number box shows how the number **325** is made up by

3 hundreds + **2** tens + **5** ones

Copy and complete the number boxes.
Write them as the one above in HUNDREDS, TENS and ONES.

1.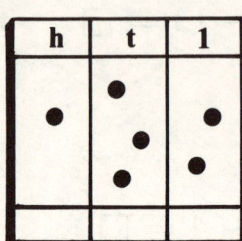

2.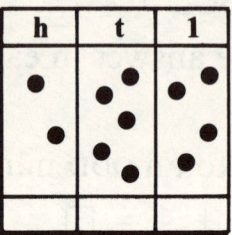

3.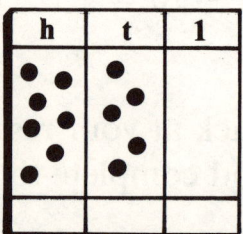

4.

B

Copy the sums and fill in the empty squares.

1. 264 = ☐ hundreds   ☐ tens   ☐ ones
2. 502 = ☐ hundreds   ☐ tens   ☐ ones
3. 625 = ☐ hundreds   ☐ tens   ☐ ones
4. 754 = ☐ hundreds   ☐ tens   ☐ ones
5. 590 = ☐ hundreds   ☐ tens   ☐ ones
6. 621 = ☐ hundreds   ☐ tens   ☐ ones
7. 520 = ☐ hundreds   ☐ tens   ☐ ones
8. 750 = ☐ hundreds   ☐ tens   ☐ ones

C

The next 5 numbers after 47 are 48, 49, 50, 51, 52.

1. Write the next 5 numbers after 69.
2. Write the next 5 numbers after 97.
3. Write the next 5 numbers *before* 203.
4. Write the next 5 numbers *before* 502.

D Five children made these scores on different pin tables.

	Alan	Beryl	Clare	David	Eric	Fiona
1st table	304	621	475	291	562	399
2nd table	428	529	243	347	675	264
3rd table	605	350	506	441	573	433
4th table	328	571	640	308	382	728

1. Put each child's scores in order, highest first.
   Alan – 605, 428, 328, 304
2. Show the order the children scored at each table, highest first.
   1st table – Beryl 621, Eric 562, Clare 475, Fiona 399, ...
3. Who had the highest score?
   At which table?
4. Who scored the least?
   At which table?

E

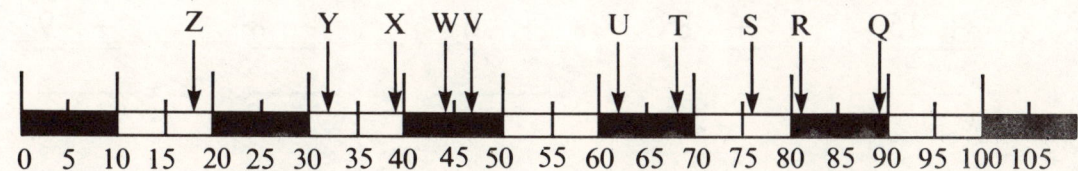

Copy the number line and the arrows marked on it.

1. Write the numbers you think the arrows are pointing to.
   This is the first: **Z → 18**
2. Is (a) 18 nearer 20 or 10? (b) 32 nearer 30 or 40?
   (c) 39 nearer 30 or 40? (d) 44 nearer 40 or 50?
   (e) 47 nearer 40 or 50? (f) 62 nearer 60 or 70?
   (g) 68 nearer 60 or 70? (h) 81 nearer 80 or 90?
   (i) 89 nearer 80 or 90? (j) 96 nearer 90 or 100?
   (k) 98 nearer 90 or 100? (l) 107 nearer 100 or 110?

Adding practice

A   Look at these adding sums.

$$
\begin{array}{r}
8 \\
+\ 7 \\
\hline
15
\end{array}
\qquad
\begin{array}{r}
12 \\
+\ 5 \\
\hline
17
\end{array}
\qquad
\begin{array}{r}
9 \\
+\ 0 \\
\hline
9
\end{array}
\qquad
\begin{array}{r}
3 \\
+\ 14 \\
\hline
17
\end{array}
$$

Find these totals.   Copy the sums neatly.

1. $\begin{array}{r}4 \\ +5 \\ \hline\end{array}$	2. $\begin{array}{r}6 \\ +6 \\ \hline\end{array}$	3. $\begin{array}{r}5 \\ +8 \\ \hline\end{array}$	4. $\begin{array}{r}4 \\ +7 \\ \hline\end{array}$	5. $\begin{array}{r}3 \\ +9 \\ \hline\end{array}$
6. $\begin{array}{r}8 \\ +4 \\ \hline\end{array}$	7. $\begin{array}{r}14 \\ +5 \\ \hline\end{array}$	8. $\begin{array}{r}15 \\ +3 \\ \hline\end{array}$	9. $\begin{array}{r}19 \\ +2 \\ \hline\end{array}$	10. $\begin{array}{r}18 \\ +3 \\ \hline\end{array}$
11. $\begin{array}{r}17 \\ +3 \\ \hline\end{array}$	12. $\begin{array}{r}4 \\ +16 \\ \hline\end{array}$	13. $\begin{array}{r}17 \\ +5 \\ \hline\end{array}$	14. $\begin{array}{r}3 \\ +19 \\ \hline\end{array}$	15. $\begin{array}{r}16 \\ +5 \\ \hline\end{array}$
16. $\begin{array}{r}5 \\ +16 \\ \hline\end{array}$	17. $\begin{array}{r}15 \\ +7 \\ \hline\end{array}$	18. $\begin{array}{r}9 \\ +17 \\ \hline\end{array}$	19. $\begin{array}{r}18 \\ +0 \\ \hline\end{array}$	20. $\begin{array}{r}0 \\ +15 \\ \hline\end{array}$
21. $\begin{array}{r}9 \\ +15 \\ \hline\end{array}$	22. $\begin{array}{r}18 \\ +8 \\ \hline\end{array}$	23. $\begin{array}{r}17 \\ +6 \\ \hline\end{array}$	24. $\begin{array}{r}19 \\ +8 \\ \hline\end{array}$	25. $\begin{array}{r}16 \\ +9 \\ \hline\end{array}$

B   Copy and find the sum of:

1. $8 + 4$	2. $18 + 4$	3. $7 + 7$
4. $17 + 7$	5. $6 + 8$	6. $16 + 8$
7. $9 + 9$	8. $19 + 9$	9. $5 + 7$
10. $5 + 17$	11. $4 + 4$	12. $4 + 14$
13. $3 + 7$	14. $13 + 7$	15. $0 + 16$

C Copy and complete the arrow graphs.
The arrow ——→ tells you what to do.

1.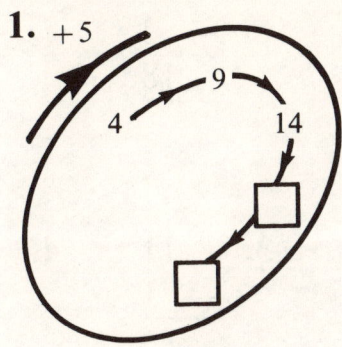

2.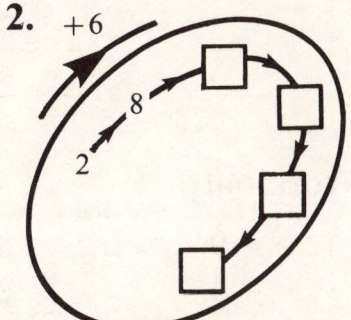

3.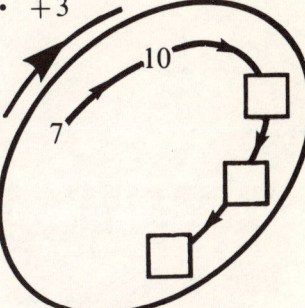

4.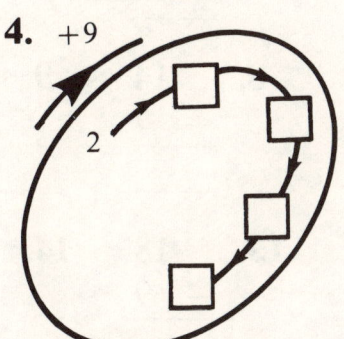

5.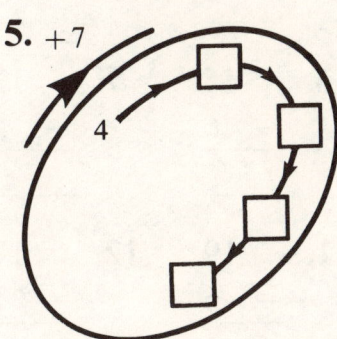

6.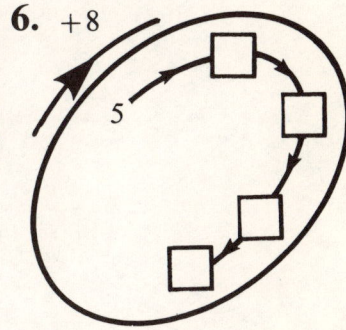

D Find these totals.

1. $2 + 3 + 5 = \square$
2. $1 + 4 + 6 = \square$
3. $1 + 8 + 7 = \square$
4. $4 + 3 + 9 = \square$
5. $3 + 3 + 3 = \square$
6. $4 + 4 + 8 = \square$
7. $4 + 8 + 4 = \square$
8. $5 + 5 + 5 = \square$
9. $7 + 6 + 2 = \square$
10. $5 + 0 + 7 = \square$
11. $6 + 9 + 6 = \square$
12. $9 + 8 + 7 = \square$
13. $7 + 8 + 7 = \square$
14. $9 + 9 + 9 = \square$
15. $6 + 4 + 5 = \square$
16. $8 + 0 + 6 = \square$
17. $5 + 9 + 0 = \square$
18. $7 + 9 + 4 = \square$
19. $6 + 8 + 7 = \square$
20. $9 + 7 + 6 = \square$
21. $6 + 8 + 1 = \square$

E Copy and complete the table of goals scored.

TEAM	home goals	away goals	TOTAL
Arsenal	15	7	**22**
Ipswich	12	9	
Liverpool	18	10	
Manchester U	17	9	
Leeds	19	8	

Subtracting

A   Look at these subtracting sums.

8	12	21	15
− 3	− 8	− 3	− 0
5	4	18	15

Work out:

1.   9
      − 4

2.   8
      − 2

3.   11
      − 1

4.   12
      − 3

5.   13
      − 3

6.   15
      − 6

7.   13
      − 5

8.   14
      − 7

9.   16
      − 9

10.   14
       − 6

11.   19
       − 6

12.   17
       − 7

13.   18
       − 9

14.   16
       − 8

15.   21
       − 2

16.   20
       − 0

17.   21
       − 6

18.   23
       − 4

19.   21
       − 5

20.   25
       − 5

B Take away:

1. 5 from 9   **2.** 7 from 11   **3.** 8 from 15
4. 0 from 10   **5.** 3 from 12   **6.** 7 from 13
7. 9 from 17   **8.** 8 from 15   **9.** 7 from 21
10. 4 from 10   **11.** 6 from 20   **12.** 9 from 22.

C   What must I add to

1. 7 to make 9   **2.** 5 to make 10   **3.** 5 to make 12
4. 4 to make 15   **5.** 4 to make 20   **6.** 8 to make 13?

D   Find the difference between 5 and 12.
5 from 12 is   12 − 5 = 7.

I must take the smaller one from the bigger. 12 is bigger so: 12 − 5

Find the difference between

1. 4 and 9	**2.** 9 and 7	**3.** 11 and 7	**4.** 3 and 7
5. 8 and 4	**6.** 5 and 9	**7.** 6 and 12	**8.** 13 and 8
9. 0 and 16	**10.** 12 and 0	**11.** 21 and 4	**12.** 19 and 9
13. 23 and 7	**14.** 24 and 9	**15.** 20 and 9	**16.** 8 and 23
17. 24 and 5	**18.** 21 and 5	**19.** 9 and 18	**20.** 17 and 8
21. 26 and 8	**22.** 25 and 7	**23.** 4 and 15	**24.** 5 and 22.

E   Copy and complete this table of goal differences.

TEAM	goals for	goals against	GOAL DIFFERENCE
Bolton	21	9	**12**
Norwich	16	7	
Spurs	14	8	
Everton	20	6	
Wolves	13	13	
C Palace	6	20	

Which team was doing worst?

F   Copy and complete these mappings.
The arrow ———→——— tells you what to do.

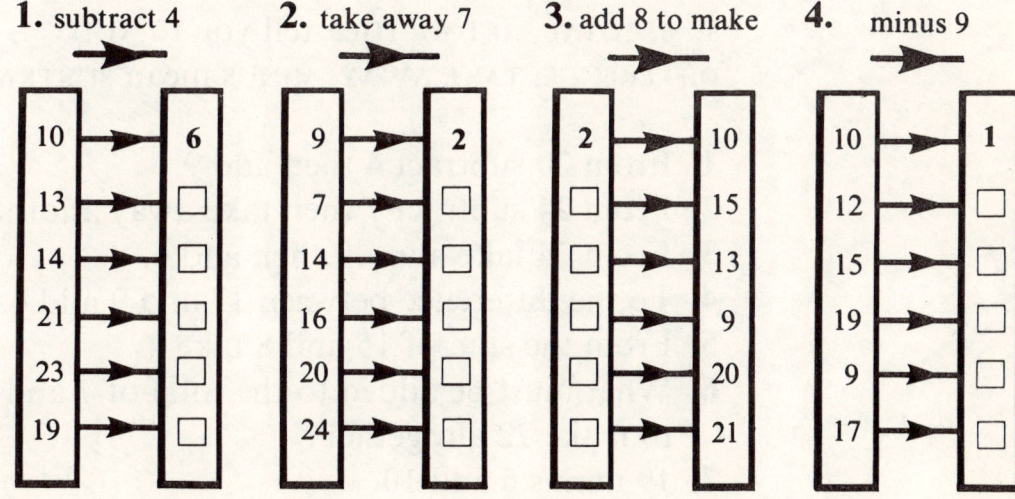

1. subtract 4   **2.** take away 7   **3.** add 8 to make   **4.** minus 9

Adding and subtracting

A   Work out these sums (the first is done for you).

1. $5 + 2 - 3 = 7 - 3 = 4$
2. $8 + 2 - 4$ 3. $7 + 9 - 3$ 4. $8 + 4 - 2$ 5. $6 + 7 - 4$
6. $5 + 9 - 2$ 7. $8 + 9 - 6$ 8. $9 + 9 - 8$ 9. $7 + 3 - 5$
10. $8 + 0 - 4$ 11. $9 + 6 - 7$ 12. $8 + 7 - 9$ 13. $9 + 8 - 9$
14. $9 + 6 - 8$ 15. $8 + 8 - 9$ 16. $7 + 9 - 8$ 17. $5 + 6 - 7$

B   Copy and complete these mappings.
Be sure to look at the instructions over the arrows.

1.
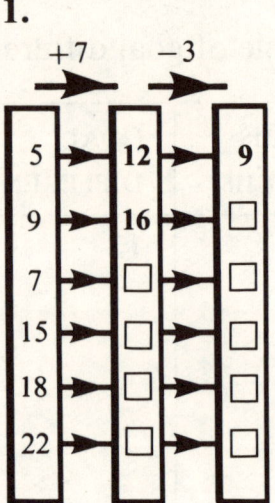

2.
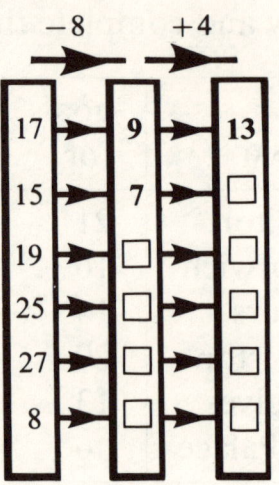

3.
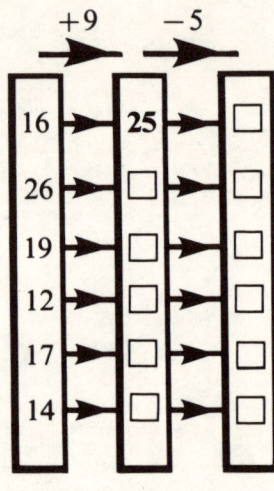

C   Read each sentence carefully.
The words tell you what to do.

SUM, TOTAL, ALTOGETHER tell you to ADD   +

DIFFERENCE, TAKE AWAY, MINUS mean SUBTRACT   −

1. From 20 subtract 4 then add 9.
2. From 24 subtract 7 then take away another 7.
3. From 24 take away 9 then add 6.
4. To the difference between 17 and 9 add 6.
5. From the sum of 15 and 8 take 4.
6. What must be added to the total of 7 and 6
    to make 22 altogether?
7. 19 minus 6 add 10.

14

D *Example:* $(7 + 4) + 3$.
Always work out *brackets* first:
$(7 + 4) + 3 = 11 + 3 = 14.$

Work out:

1. $(5 + 4) + 3$	**2.** $5 + (4 + 3)$	**3.** $(6 + 3) + 7$	**4.** $6 + (3 + 7)$
5. $(9 + 8) + 2$	**6.** $9 + (8 + 2)$	**7.** $(7 + 6) + 3$	**8.** $7 + (6 + 3)$
9. $(12 + 7) + 3$	**10.** $12 + (7 + 3)$	**11.** $(15 + 5) + 9$	**12.** $15 + (5 + 9)$
13. $(14 + 6) + 4$	**14.** $14 + (6 + 4)$	**15.** $(17 + 5) + 6$	**16.** $17 + (5 + 6)$
17. $(5 + 9) + 11$	**18.** $5 + (9 + 11)$	**19.** $(7 + 8) + 12$	**20.** $7 + (8 + 12)$

Did you get the same answer for each pair of sums?

E Copy and complete these. (Be sure to work out *brackets* first.)

1. $(7 + 6) - 2$	**2.** $7 + (6 - 2)$	**3.** $(8 + 9) - 7$	**4.** $8 + (9 - 7)$
5. $(4 + 11) - 8$	**6.** $4 + (11 - 8)$	**7.** $(14 - 7) + 3$	**8.** $14 - (7 + 3)$
9. $(18 - 9) + 6$	**10.** $18 - (9 + 6)$	**11.** $20 - (8 + 9)$	**12.** $(20 - 8) + 9$
13. $(21 - 6) - 4$	**14.** $21 - (6 - 4)$	**15.** $(18 - 7) - 3$	**16.** $18 - (7 - 3)$

Was each pair of answers the same this time?

F Copy and complete the cross number puzzle.
To fill in the squares work out the answers to the clues on each
side of the puzzle.

Clues Across

1 $(6 + 8) - 2$
3 $(20 + 7) - 1$
4 $(14 + 6) + 5$
5 $22 - (2 + 3)$
7 $(6 + 7) + 10$
8 $(19 + 10) - 1$
9 $18 - (7 - 4)$

Clues Down

1 $(12 + 8) - 4$
2 $20 - (3 + 2)$
3 $(6 + 7) + 8$
5 $(12 + 7) - 1$
6 $5 + (6 + 7)$
7 $(17 + 7) - 3$
8 $(19 + 8) - 2$

Measuring – length

A

Measure the lines carefully. Draw them in centimetres (cm).
Write each length at the side.

———————————————— t = 7 cm

———————————— u = ☐ cm

————————————————— v = ☐ cm

——————————————— w = ☐ cm

——————————————————— x = ☐ cm

—————————— y = ☐ cm

————————————————— z = ☐ cm

1. Which line is the longest?
2. Which line is the shortest?

B Use your answers in Section A to work out:

1. t + u	**2.** t + v	**3.** t + w	**4.** t + x	**5.** t + y
6. t + z	**7.** u + v	**8.** u + w	**9.** u + x	**10.** u + y
11. u + z	**12.** v + w	**13.** v + z	**14.** w + x	**15.** x + y
16. x + z	**17.** y + z	**18.** x + y + z	**19.** t + u + v	**20.** v + w + y

C Find these sums.

1. cm	**2.** cm	**3.** cm	**4.** cm	**5.** cm	**6.** cm
5	8	5	9	8	17
+ 9	+ 6	+ 7	+ 8	+ 7	+ 4

7. cm	**8.** cm	**9.** cm	**10.** cm	**11.** cm	**12.** cm
15	16	17	12	18	19
+ 7	+ 8	+ 6	+ 9	+ 8	+ 9

13. 18 cm + 6 cm   **14.** 15 cm + 5 cm   **15.** 19 cm + 10 cm

D   Measure these lines in centimetres (cm). Copy them carefully.

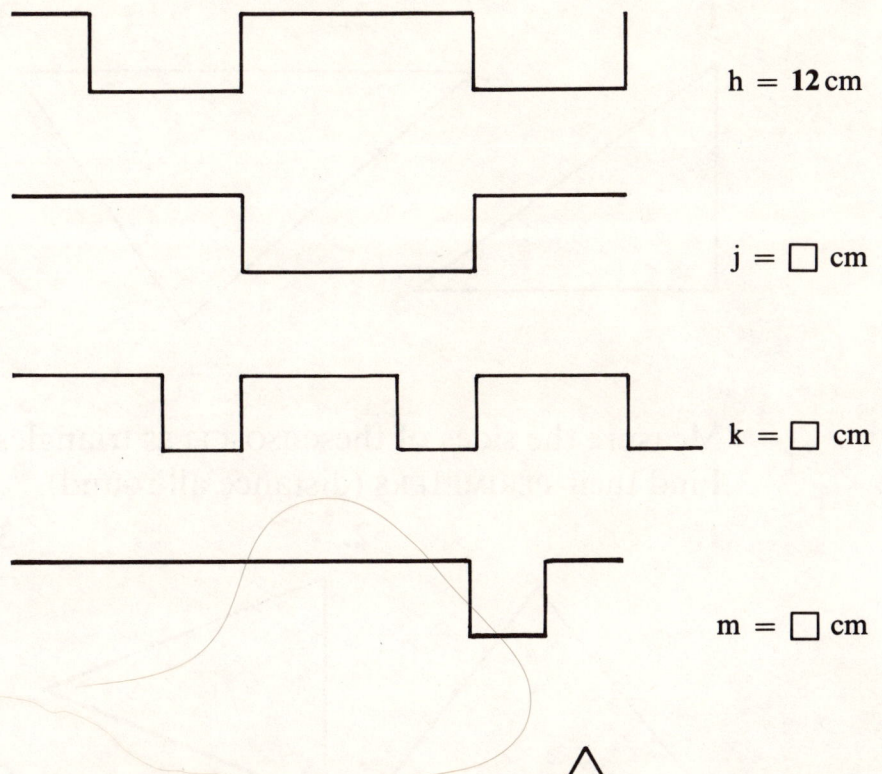

h = **12** cm

j = ☐ cm

k = ☐ cm

m = ☐ cm

n = ☐ cm

E   Use your lengths in Section **D**. How much longer is:
1. h than j   2. h than m   3. h than n   4. k than h
5. m than n   6. j than n   7. k than j   8. k than m?

F   Find the difference in length between:
1. m and h   2. n and k   3. k and n   4. j and n.

G   Work out these differences.

1. cm	2. cm	3. cm	4. cm	5. cm
9	8	12	13	15
− 2	− 5	− 7	− 9	− 8

6. 21 cm − 5 cm   **7.** 19 cm − 10 cm   **8.** 24 cm − 8 cm

Perimeters – triangles

A   Measure the sides of these triangles in centimetres (cm).
Find their PERIMETERS (distance all round).

1.   **2.**   **3.**

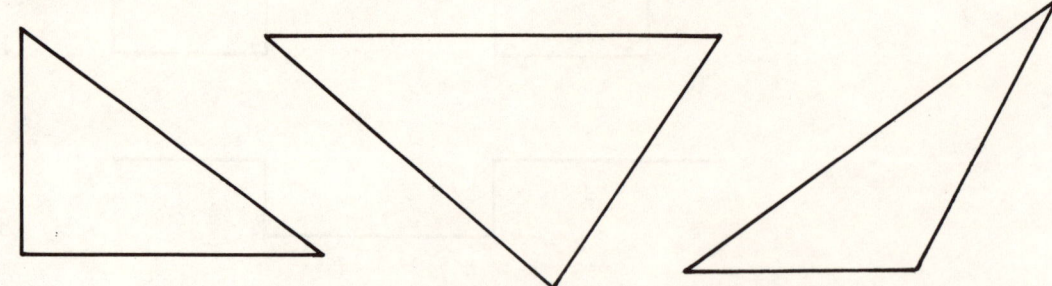

B   Measure the sides of these ISOSCELES triangles in centimetres (cm).
Find their PERIMETERS (distance all round).

1.   **2.**   **3.**

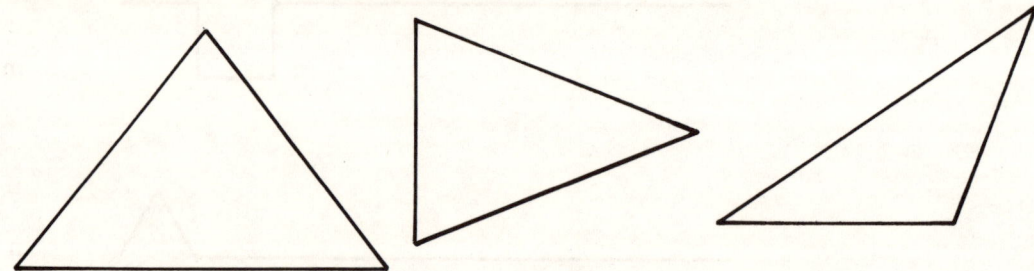

What is special about ISOSCELES triangles?

C   Measure the sides of these EQUILATERAL triangles in centimetres (cm).
Find their PERIMETERS (distance all round).

1.   **2.**   **3.**

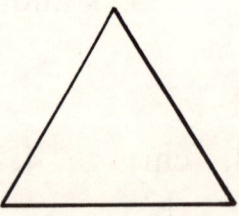 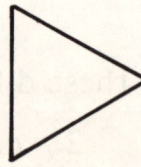 

What is special about EQUILATERAL triangles?

D Measure each side of these QUADRILATERALS.
Find their PERIMETERS (distance all round) in centimetres (cm).

1.   **2.**   **3.**

E Measure each side of these RECTANGLES.
Work out their PERIMETERS in centimetres (cm).

1.   **2.**   **3.**

A RECTANGLE is a QUADRILATERAL.
What is special about a RECTANGLE?

F Work out the PERIMETERS of these SQUARES in centimetres (cm).

1.   **2.**

What is special about a SQUARE?

G **1.** How many sides has any QUADRILATERAL?
2. How many sides has any TRIANGLE?

Pence

A How much are these sets of coins worth?

1.

$10p + 2p + 5p = \square$

2.

$5p + 2p + 1p = \square$

3.

$5p + 5p + 2p = \square$

4.

$10p + 10p + 2p = \square$

5.

$20p + 10p + 5p = \square$

6.

$20p + 5p + 2p + 2p = \square$

B Which coins would you use to make:

1. 7p	**2.** 9p	**3.** 4p	**4.** 11p	**5.** 13p
6. 12p	**7.** 15p	**8.** 14p	**9.** 18p	**10.** 19p
11. 21p	**12.** 23p	**13.** 17p	**14.** 28p	**15.** 26p?

C 1. To make 10p you need: (a) $\square$ 2p coins (b) $\square$ 5p coins.

2. To make 20p you need: (a) $\square$ 2p coins (b) $\square$ 5p coins (c) $\square$ 10p coins.

3. To make 30p you need: (a) $\square$ 2p coins (b) $\square$ 5p coins (c) $\square$ 10p coins.

D How much would I have if I had these coins:

1. two 5p and three 1p?   2. two 10p and one 5p?
3. three 2p and three 1p?   4. two 2p and two 5p?
5. four 2p and one 5p?   6. one 20p and three 2p?

E Work out these totals.

1. p	2. p	3. p	4. p	5. p
7	10	9	8	6
+ 6	+ 7	+ 5	+ 7	+ 9

6. p	7. p	8. p	9. p	10. p
15	13	17	12	18
+ 4	+ 7	+ 4	+ 9	+ 6

11. p	12. p	13. p	14. p	15. p
3	7	12	7	8
5	4	5	13	8
+ 4	+ 5	+ 6	+ 2	+ 8

16. 4p + 4p + 4p   17. 5p + 6p + 7p   18. 7p + 10p + 3p

F Find the change from 20p after spending:

1. 16p	2. 19p	3. 15p
4. 17p	5. 12p	6. 8p
7. 5p	8. 9p	9. 6p
10. 2p	11. 10p	12. 3p.

G What coins would you receive in change in Section F?
The shopkeeper gives you the fewest coins in each case.

Time

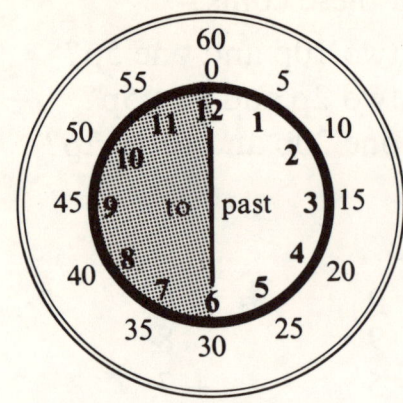

a.m. means 'in the morning'.
p.m. means 'in the afternoon' (or evening).
5 minutes past 8 in the morning is written **8.05 a.m.**
5 minutes past 8 in the evening is written **8.05 p.m.**

Write these times using figures and a.m. or p.m.

1. 5 minutes past 10 in the morning
2. 5 minutes past 10 in the evening
3. 10 minutes past 6 in the afternoon
4. 25 minutes past 4 in the afternoon
5. quarter past 6 in the morning
6. half past 2 in the afternoon
7. 25 minutes to 9 in the morning
8. 5 minutes to 12 in the evening
9. 10 minutes to 8 in the evening
10. quarter to 9 in the morning
11. 20 minutes to 11 in the morning
12. half past 5 in the afternoon

NOON is 12 o'clock or 12.00 MIDDAY.
MIDNIGHT is also 12 o'clock or 12.00.

B 1.   **2.**   **3.**   **4.**   **5.**

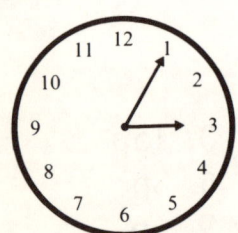 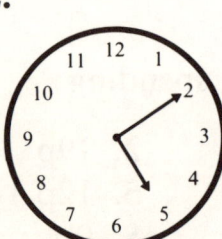 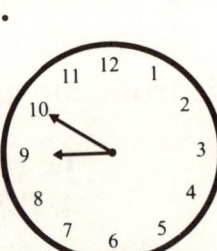 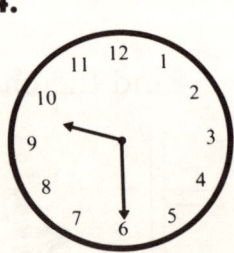 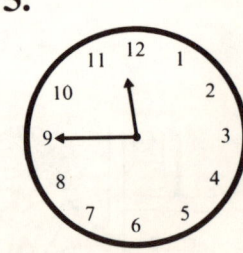

Write, in figures, the times on these clocks.

C

Look at the clock times in Section B, 1–5, above.
Draw clocks to show the *correct* times if the clocks in Section B
are:
(a) 5 minutes slow   (b) 10 minutes fast.
Record, in figures, the new times on your clocks.

D 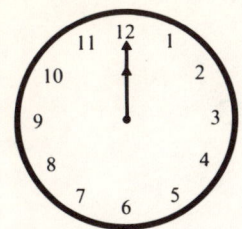 This clock is showing the time as NOON, 12.00.
Write the new times when the clock has moved on:

1. 5 minutes
2. 15 minutes
3. 20 minutes
4. half an hour
5. 35 minutes
6. 40 minutes
7. 45 minutes
8. 50 minutes
9. 55 minutes
10. 60 minutes, or 1 hour.

E How many minutes are there from:

1. 1.15 p.m. to 1.20 p.m.
2. 8.30 a.m. to 8.40 a.m.
3. 4.13 a.m. to 4.20 a.m.
4. 6.25 p.m. to 6.35 p.m.
5. 11.45 a.m. to 11.50 a.m.
6. 11.55 a.m. to NOON
7. 11.55 a.m. to 12.05 p.m.
8. 11.55 p.m. to 12.10 a.m.?

F Below is the timetable for the school bus.

	TIME
bus sets off	8.22
picks up Alan	8.28
picks up Brian	8.33
picks up Claire	8.40
picks up Diane	8.42
picks up Edith	8.51
arrives at school	8.55

Work out the number of minutes between:

1. setting out and picking up Alan.
2. picking up Alan and picking up Brian.
3. picking up Brian and picking up Claire.
4. picking up Claire and picking up Diane.
5. picking up Diane and picking up Edith.
6. picking up Edith and arriving at school.
7. setting out and reaching school.

Adding tens and ones

A *Examples:*

1.
$$39 \rightarrow (30 + 9)$$
$$\underline{+ 4} \rightarrow \underline{+ 4}$$
$$30 + 13 \rightarrow 30 + 10 + 3 \rightarrow 40 + 3 \rightarrow 43.$$

2.
$$74 \rightarrow (70 + 4)$$
$$\underline{+ 8} \rightarrow \underline{+ 8}$$
$$70 + 12 \rightarrow 70 + 10 + 2 \rightarrow 80 + 2 \rightarrow 82.$$

Add these:

1. 49 + 9	**2.** 67 + 6	**3.** 75 + 9	**4.** 32 + 9	**5.** 23 + 8
6. 39 + 9	**7.** 87 + 4	**8.** 89 + 8	**9.** 59 + 6	**10.** 76 + 7
11. 88 + 8	**12.** 84 + 9	**13.** 46 + 6	**14.** 38 + 8	**15.** 54 + 9

B Add these 10's:

1. 50 + 20	**2.** 40 + 50	**3.** 70 + 10	**4.** 30 + 40	**5.** 30 + 60
6. 40 + 40	**7.** 10 + 60	**8.** 90 + 10	**9.** 80 + 20	**10.** 70 + 30

11. 20 + 30 + 40   **12.** 10 + 30 + 20   **13.** 30 + 30 + 30

C *Examples:*

1. $32 \rightarrow \quad (30 + 2)$
    $+\ 28 \rightarrow +\ (20 + 8)$
    $\overline{ \quad \overline{50 + 10} \rightarrow 60.}$

2. $35 \rightarrow \quad (30 + 5)$
    $+\ 27 \rightarrow +\ (20 + 7)$
    $\overline{ \quad \overline{50 + 12} \rightarrow 50 + 10 + 2 \rightarrow 62.}$

Find the sum of:

1. 16 $+\ 14$	**2.** 41 $+\ 29$	**3.** 22 $+\ 38$	**4.** 36 $+\ 25$	**5.** 24 $+\ 48$
6. 29 $+\ 18$	**7.** 57 $+\ 37$	**8.** 58 $+\ 33$	**9.** 65 $+\ 25$	**10.** 23 $+\ 67$
11. 35 $+\ 47$	**12.** 19 $+\ 33$	**13.** 26 $+\ 49$	**14.** 39 $+\ 59$	**15.** 54 $+\ 45$

16. 25 + 15 **17.** 34 + 17 **18.** 48 + 16 **19.** 59 + 26

D This table shows the points scored
at an athletics meeting.
Complete the totals.

COUNTRY	MEN	WOMEN	TOTAL
France	49	37	
Italy	59	26	
Germany	63	29	
Britain	57	34	
Belgium	44	27	
Holland	39	39	

Subtracting tens and ones

A *Example:*

$$98 \longrightarrow (90 + 8)$$
$$- 46 \rightarrow -(40 + 6)$$
$$50 + 2 \rightarrow 52.$$

Work out:

1. 87 − 26	**2.** 59 − 37	**3.** 56 − 34	**4.** 84 − 31	**5.** 75 − 23
6. 94 − 21	**7.** 87 − 54	**8.** 59 − 27	**9.** 64 − 61	**10.** 89 − 82

B *Example:*

$$74 \longrightarrow (70 + 4) \quad \rightarrow (60 + 14)$$
$$- 39 \rightarrow -(30 + 9) \rightarrow -(30 + \ 9)$$
$$30 + \ 5 \rightarrow 35.$$

Try these

1. 75 − 28	**2.** 71 − 59	**3.** 31 − 27	**4.** 54 − 28	**5.** 67 − 48
6. 81 − 18	**7.** 83 − 66	**8.** 82 − 68	**9.** 56 − 38	**10.** 30 − 19
11. 90 − 42	**12.** 85 − 57	**13.** 91 − 19	**14.** 62 − 26	**15.** 51 − 15

16. 77 − 29   **17.** 52 − 39   **18.** 64 − 47   **19.** 40 − 27

C

1. Copy and complete this table. It shows the children in Chantry School.

	1st year		2nd year		3rd year		4th year	
CLASS	1	2	3	4	5	6	7	8
boys girls	14 17	13 16	17 17	16	17	15	13	16
TOTAL	**31**			31	32	28	29	33

2. How many children are in each year altogether?
3. How many boys are in each year?
4. How many girls are in each year?

D

1. Find the total number of points in this hand of cards.
2. The 9 and 7 cards are played. How many points are left?
3. How many points must then be picked up to make 21 altogether?
4. How can this be done using not more than two cards?

E

A bus can hold 63 passengers.
28 seats are empty.
1. How many passengers are there?
2. 19 passengers get on. How many seats are then empty?
3. How many passengers are there altogether then?

F

Janet and John swam in a race.
Janet took 36 seconds. John took 51 seconds.
1. Who won the race?
2. By how many seconds was the race won?

G

A candle is 30 cm long when new.
1. How long will it be after 2 cm have burned away?
2. How long will it be after 12 cm have burned away?
3. How long will it be after 19 cm have burned away?

Adding centimetres

A Work out:

1. cm	**2.** cm	**3.** cm	**4.** cm	**5.** cm
19	21	35	37	46
+ 12	+ 17	+ 21	+ 15	+ 26

6. cm	**7.** cm	**8.** cm	**9.** cm	**10.** cm
54	48	63	43	37
+ 27	+ 42	+ 29	+ 48	+ 46

11. 78 cm + 9 cm   **12.** 27 cm + 35 cm   **13.** 77 cm + 19 cm
14. 6 cm + 55 cm   **15.** 38 cm + 29 cm   **16.** 54 cm + 26 cm
17. 26 cm + 26 cm   **18.** 45 cm + 25 cm   **19.** 62 cm + 28 cm
20. 89 cm + 10 cm   **21.** 37 cm + 37 cm   **22.** 56 cm + 30 cm

B

1. The length of a rectangle is 27 cm. Its width is 19 cm.
    (a) Find the sum of the length and the width.
    (b) Work out the perimeter of the rectangle (distance all round).

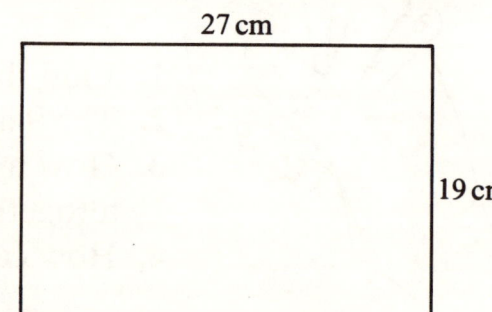

27 cm

19 cm

2. A baby was 57 cm tall. How tall was she after growing
    (a) 5 cm   (b) 12 cm   (c) 14 cm   (d) 19 cm?

3. A brick is 7 cm thick. How high will a pile be made up of
    (a) 2 bricks   (b) 3 bricks   (c) 4 bricks   (d) 5 bricks?

4. A line is 45 cm long. How long is it after being increased by 29 cm?

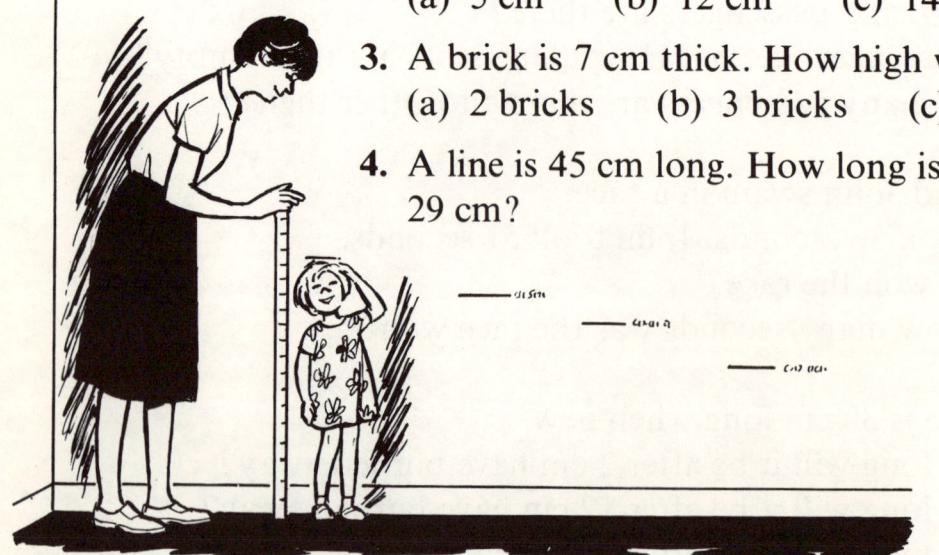

C   Work out the PERIMETERS (distance all round) of these triangles.

1.   **2.**   **3.**

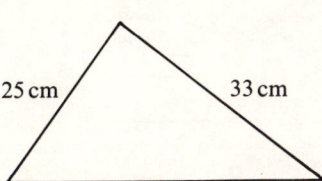
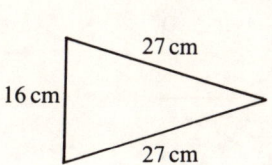
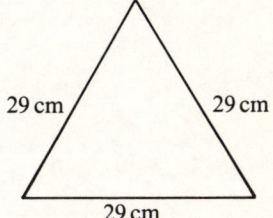

4. Which triangle is ISOSCELES?

5. Which triangle is EQUILATERAL?

D   Work out the PERIMETERS of these QUADRILATERALS.

1.   **2.**   **3.**

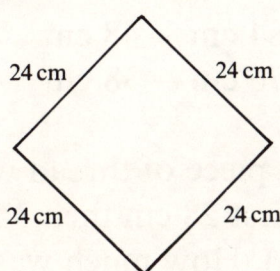
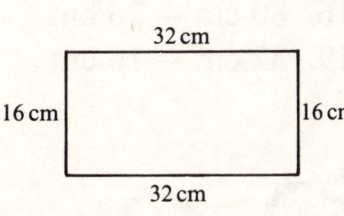
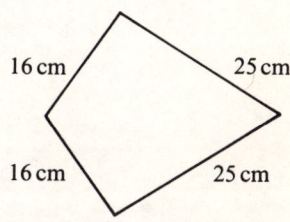

4.   **5.**   **6.**

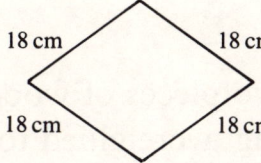
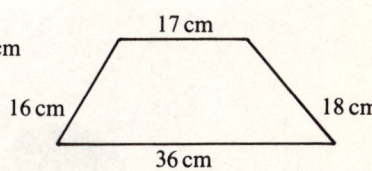

7. Which QUADRILATERAL is a SQUARE?

8. Which QUADRILATERAL is a RECTANGLE?

All the other QUADRILATERALS have special names.
Try to find out what they are called.

Subtracting centimetres

A Work out:

1. cm	**2.** cm	**3.** cm	**4.** cm	**5.** cm
24	26	23	24	32
− 9	− 13	− 17	− 16	− 18

6. cm	**7.** cm	**8.** cm	**9.** cm	**10.** cm
35	32	39	43	54
− 19	− 26	− 12	− 18	− 28

11. cm	**12.** cm	**13.** cm	**14.** cm	**15.** cm
62	74	56	92	95
− 44	− 37	− 28	− 46	− 35

16. 80 cm − 26 cm   **17.** 84 cm − 48 cm   **18.** 52 cm − 26 cm
19. 32 cm − 16 cm   **20.** 76 cm − 38 cm   **21.** 91 cm − 19 cm

B

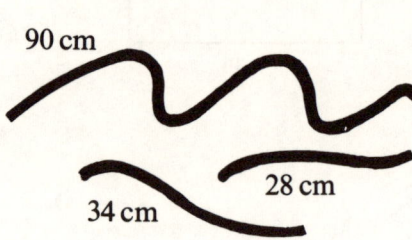

1. A piece of thread was 90 cm long.
   First 28 cm then 34 cm were cut off.
   (a) How much was left after the first
       cut?
   (b) How much was left after the second
       cut?
   (c) How much was cut off altogether?

2. Two pieces of wood, 48 cm and 25 cm
   long, were glued together.
   (a) What was their total length?
   (b) They overlapped 13 cm after glueing.
       What was the final length?

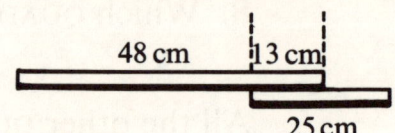

C   Work out the odd side of these triangles.
The first is done for you.

1.

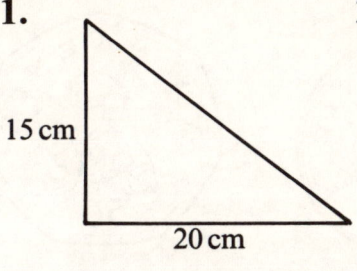

perimeter 60 cm

**sum of the 2 sides
is (20 + 15) cm
odd side is
60 − (20 + 15) cm
= (60 − 35) cm
= 25 cm**

2.

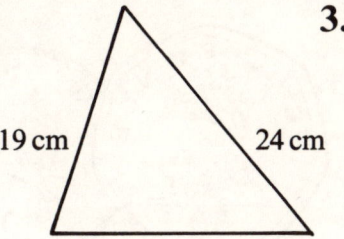

perimeter 68 cm

3.

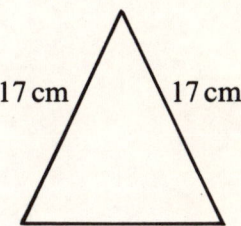

perimeter 48 cm

4.

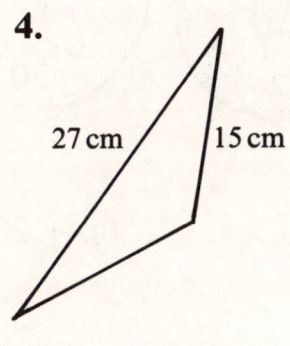

perimeter 57 cm

5.

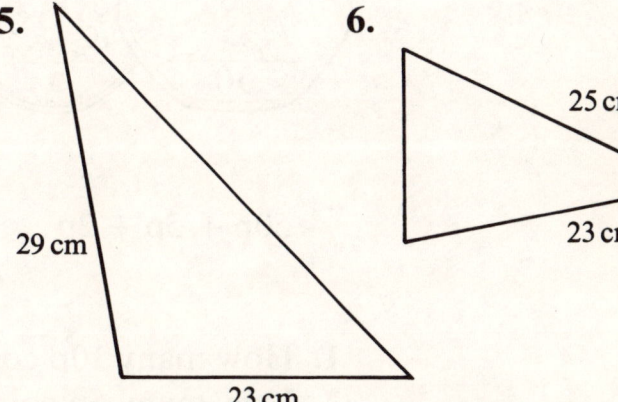

perimeter 96 cm

6.

perimeter 63 cm

7. Which of these triangles are ISOSCELES TRIANGLES?

D   **1.** A rectangle is 44 cm long.
It is 7 cm longer than it is wide.
(a) How wide is the rectangle?
(b) What is the length and width
together?
(c) Work out the PERIMETER of
the rectangle.
2. Find the sides of a square with PERIMETER 8 cm.

44 cm

width

length

Money

A   Work out the total value of these coins.

1.

50p + 10p

2.

50p + 10p + 5p

3.

50p + 5p + 2p

4.

50p + 20p + 2p

B   **1.** How many 10p coins make 50p?
       2. How many 5p coins make 50p?

C   Which coins would you use to make:

    1. 21p   **2.** 24p   **3.** 25p   **4.** 29p   **5.** 30p
    6. 36p   **7.** 42p   **8.** 45p   **9.** 47p   **10.** 50p
   11. 52p   **12.** 56p   **13.** 59p   **14.** 61p   **15.** 67p?

D   How much would I have if I were given these coins:

    1. two 10p and a 5p   **2.** three 10p and three 2p
    3. one 50p and two 10p   **4.** one 50p and two 5p
    5. one 50p and four 2p   **6.** one 50p, one 5p and two 2p
    7. one 50p and two 20p   **8.** one 50p, one 20p and two 5p?

E   Work out these totals.

1. p	**2.** p	**3.** p	**4.** p	**5.** p
24	35	34	74	56
+ 17	+ 16	+ 9	+ 16	+ 18

6. p	**7.** p	**8.** p	**9.** p	**10.** p
54	48	29	68	35
+ 23	+ 36	+ 52	+ 24	+ 37

11. p	**12.** p	**13.** p	**14.** p	**15.** p
18	42	37	23	19
7	18	24	28	27
+ 12	+ 22	+ 26	+ 37	+ 48

16. 24p + 17p + 32p   **17.** 35p + 9p + 28p
18. 52p + 18p + 8p   **19.** 36p + 35p + 28p
20. 23p + 24p + 25p   **21.** 19p + 20p + 21p

F   Use the price-list to make out these newsagent's bills.
Copy the bills and then fill them in.

PRICE LIST	
Telegraph	18p
Mirror	12p
Sun	11p
Observer	25p
Express	15p
Mail	14p

1.
Mr Jones   p
Telegraph
Observer ___
TOTAL

2.
Mr Brown   p
Sun
Express ___
TOTAL

3.
Mr Rogers   p
Mirror
Mail ___
TOTAL

4.
Mr Collins   p
Observer
Express ___
TOTAL

More money

A   Work out:

1. p	**2.** p	**3.** p	**4.** p	**5.** p
16	17	26	23	36
− 12	− 9	− 18	− 15	− 17

6. p	**7.** p	**8.** p	**9.** p	**10.** p
42	53	84	80	57
− 27	− 29	− 39	− 36	− 28

11. p	**12.** p	**13.** p	**14.** p	**15.** p
63	71	54	94	99
− 25	− 46	− 27	− 28	− 37

16. 44p − 28p   **17.** 32p − 9p   **18.** 64p − 32p
19. 76p − 38p   **20.** 94p − 47p   **21.** 60p − 28p

B   How much more is a 50p coin worth than:
1. two 10p coins and a 5p
2. one 10p coin, one 5p and two 2p
3. three 10p coins and three 2p coins
4. one 20p coin, one 5p, two 2p and one 1p
5. three 2p coins?

C   **1.** How much change from 30p is there after spending these
amounts:
  (a) 15p   (b) 19p   (c) 22p   (d) 27p   (e) 12p?
2. How much change from 50p is there after spending these
amounts:
  (a) 12p   (b) 17p   (c) 28p   (d) 33p   (e) 41p?

D   Mrs Driver liked to use the fewest coins for her customers'
change.
She would pay 24p with one 20p coin and two 2p coins.
How would she give these amounts:

    1. 25p   **2.** 29p   **3.** 17p   **4.** 30p   **5.** 34p
    6. 41p   **7.** 45p   **8.** 56p   **9.** 62p   **10.** 65p
    11. 70p   **12.** 76p   **13.** 82p   **14.** 85p   **15.** 99p?

E   **1.** Sally bought a Crunchie bar for 15p and a comic for 17p at
one shop.
She then bought a bottle of lemonade for 25p at another shop.
(a) How much did she spend at the first shop?
(b) How much did she spend altogether?
(c) How much was left from 70p?

    2. Tom had a day at the seaside.
He spent 16p on ice-cream, 25p for a ball, 15p for candy-floss
and 19p for lemonade. How much did he spend altogether?

F Read the prices charged at the
Motorway Cafe.
Work out the prices the cashier
should charge for:

Tea	17p	Scone	18p
Coffee	22p	Pastry	29p
Milk	15p	Sausage roll	25p
Fish	62p	Chips	19p
Hamburger	37p	Pie	42p

1.   p
  fish and
    chips
    scone ____
  TOTAL ____

2.   p
  hamburger
    tea
    pastry ____
  TOTAL ____

3.   p
  pie and
    chips
    coffee
    scone ____
  TOTAL ____

4.   p
  sausage roll
    and chips
    milk ____
  TOTAL ____

5.   p
  coffee
    scone
    pastry ____
  TOTAL ____

Two times

A Copy and continue this pattern.

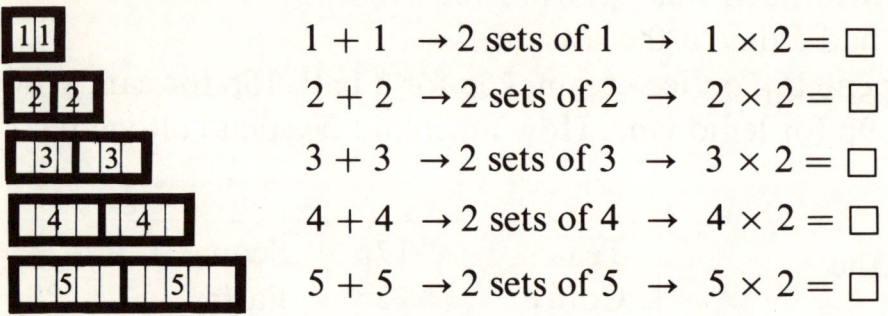

2 $\longrightarrow$ 1 set of 2 $\rightarrow$ 2 × 1 = ☐

2 + 2 $\longrightarrow$ 2 sets of 2 $\rightarrow$ 2 × 2 = ☐

2 + 2 + 2 $\longrightarrow$ 3 sets of 2 $\rightarrow$ 2 × 3 = ☐

2 + 2 + 2 + 2 $\longrightarrow$ 4 sets of 2 $\rightarrow$ 2 × 4 = ☐

2 + 2 + 2 + 2 + 2 $\rightarrow$ 5 sets of 2 $\rightarrow$ 2 × 5 = ☐

Continue up to $\longrightarrow$ 10 sets of 2 $\rightarrow$ 2 × 10 = ☐

B Copy and continue this pattern.

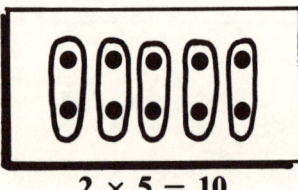

1 + 1 $\rightarrow$ 2 sets of 1 $\rightarrow$ 1 × 2 = ☐

2 + 2 $\rightarrow$ 2 sets of 2 $\rightarrow$ 2 × 2 = ☐

3 + 3 $\rightarrow$ 2 sets of 3 $\rightarrow$ 3 × 2 = ☐

4 + 4 $\rightarrow$ 2 sets of 4 $\rightarrow$ 4 × 2 = ☐

5 + 5 $\rightarrow$ 2 sets of 5 $\rightarrow$ 5 × 2 = ☐

Continue up to $\rightarrow$ 10 + 10 $\rightarrow$ 2 sets of 10 $\rightarrow$ 10 × 2 = ☐

C

The drawings show that
2 × 5 = 10
and 5 × 2 = 10.

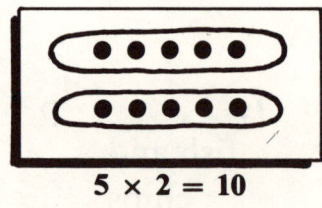

2 × 5 = 10   5 × 2 = 10

Make similar drawings for the following, and complete them.

1. 2 × 7 = **14**
7 × 2 = ☐

2. 4 × 2 = ☐
2 × 4 = ☐

3. 6 × 2 = ☐
2 × 6 = ☐

4. 8 × 2 = ☐
2 × 8 = ☐

5. 3 × 2 = ☐
2 × 3 = ☐

6. 1 × 2 = ☐
2 × 1 = ☐

7. 9 × 2 = ☐
2 × 9 = ☐

8. 0 × 2 = ☐
2 × 0 = ☐

9. 10 × 2 = ☐
2 × 10 = ☐

10. 12 × 2 = ☐
2 × 12 = ☐

11. 11 × 2 = ☐
2 × 11 = ☐

12. 2 × 2 = ☐

D Copy and complete the number line.

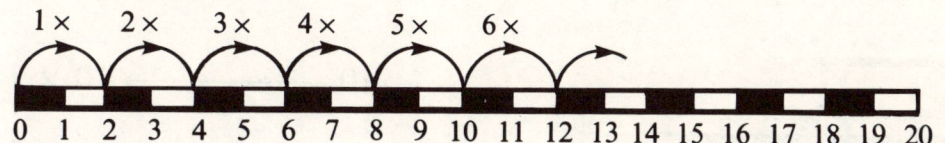

Work out:

1. $2 \times 2 = \square$	**2.** $4 \times 2 = \square$	**3.** $2 \times 5 = \square$
4. $2 \times 3 = \square$	**5.** $7 \times 2 = \square$	**6.** $2 \times 1 = \square$
7. $2 \times 8 = \square$	**8.** $10 \times 2 = \square$	**9.** $0 \times 2 = \square$
10. $2 \times 0 = \square$	**11.** $2 \times 10 = \square$	**12.** $2 \times 6 = \square$

E   Copy and complete the mappings.

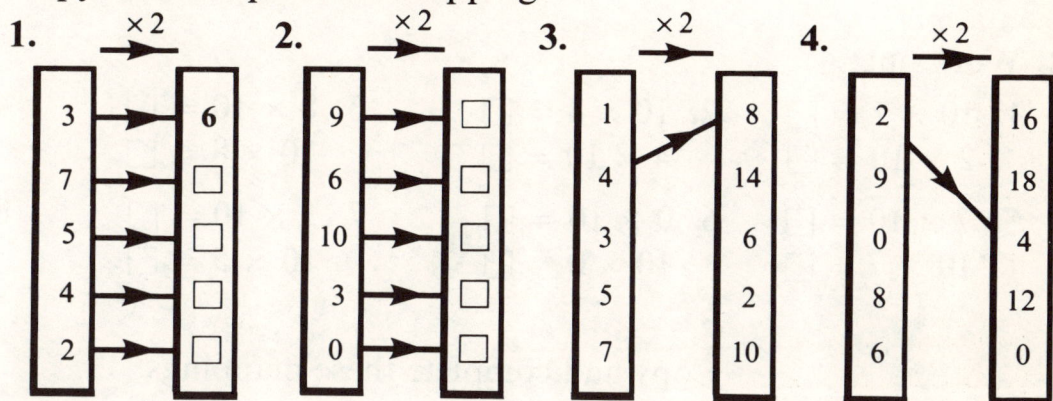

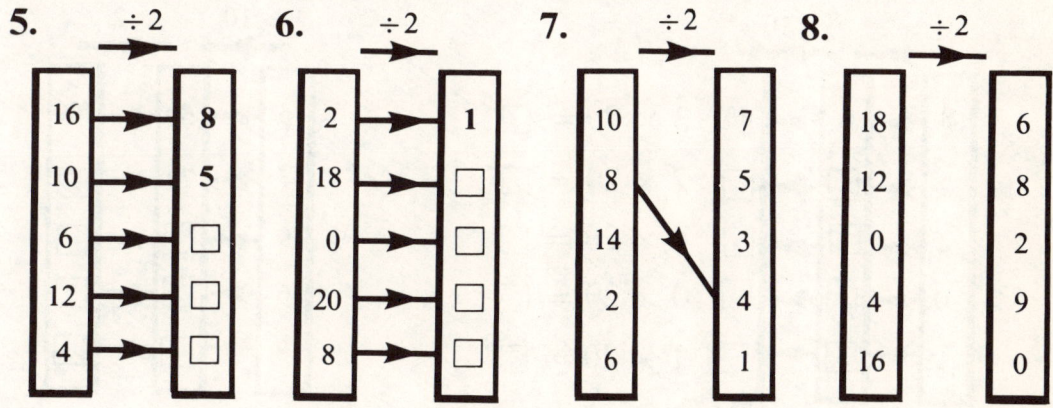

F Work out:

1. 5 $\times 2$ ___	**2.** 4 $\times 2$ ___	**3.** 10 $\times 2$ ___	**4.** 6 $\times 2$ ___	**5.** 9 $\times 2$ ___
6. 7 $\times 2$ ___	**7.** 3 $\times 2$ ___	**8.** 0 $\times 2$ ___	**9.** 8 $\times 2$ ___	**10.** 2 $\times 2$ ___

Table families

A Copy the diagram.

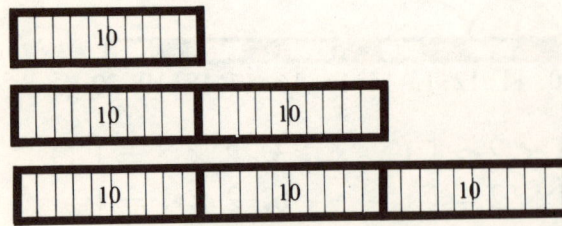

$10 \longrightarrow 10 \times 1 = 10$

$10 + 10 \longrightarrow 10 \times 2 = 20$

$10 + 10 + 10 \rightarrow 10 \times 3 = 30$

Continue up to $\longrightarrow 10 \times 10$

What do you notice when you multiply by 10?

B Work out:

1. $10 \times 2 = \square$
    $2 \times 10 = \square$

2. $10 \times 4 = \square$
    $4 \times 10 = \square$

3. $8 \times 10 = \square$
    $10 \times 8 = \square$

4. $3 \times 10 = \square$
    $10 \times 3 = \square$

5. $7 \times 10 = \square$
    $10 \times 7 = \square$

6. $0 \times 10 = \square$
    $10 \times 0 = \square$

7. $5 \times 10 = \square$
    $10 \times 5 = \square$

8. $10 \times 6 = \square$
    $6 \times 10 = \square$

C Copy and complete these mappings.

1.

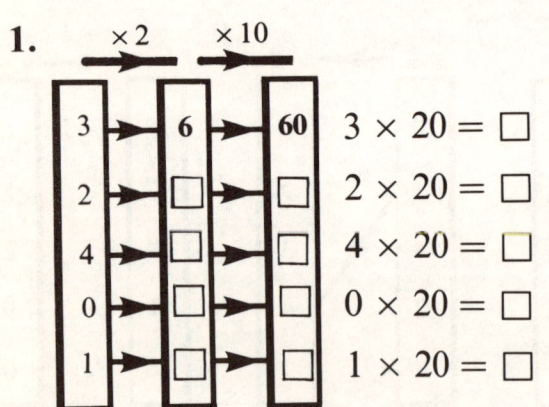

$3 \times 20 = \square$

$2 \times 20 = \square$

$4 \times 20 = \square$

$0 \times 20 = \square$

$1 \times 20 = \square$

2.

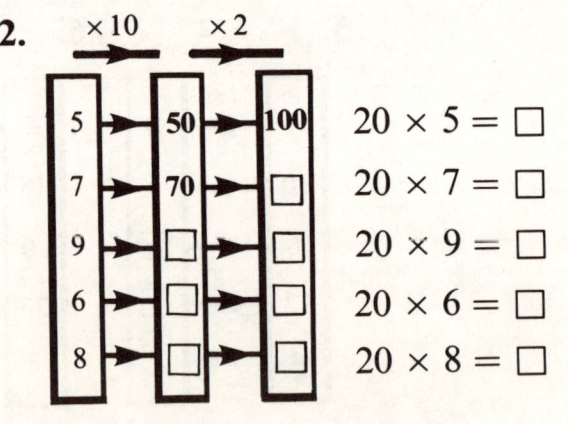

$20 \times 5 = \square$

$20 \times 7 = \square$

$20 \times 9 = \square$

$20 \times 6 = \square$

$20 \times 8 = \square$

D Work out these pairs.

1. $1 \times 10 = \square$
    $2 \times 5 = \square$

2. $2 \times 10 = \square$
    $4 \times 5 = \square$

3. $3 \times 10 = \square$
    $6 \times 5 = \square$

4. $4 \times 10 = \square$
    $8 \times 5 = \square$

5. $0 \times 10 = \square$
    $0 \times 5 = \square$

6. $10 \times 10 = \square$
    $10 \times 5 = \square$

7. $10 \times 3 = \square$
    $5 \times 3 = \square$

8. $10 \times 1 = \square$
    $1 \times 5 = \square$

E   Complete this sequence of 5's: 5, 10, 15, 20, ... 50.

F

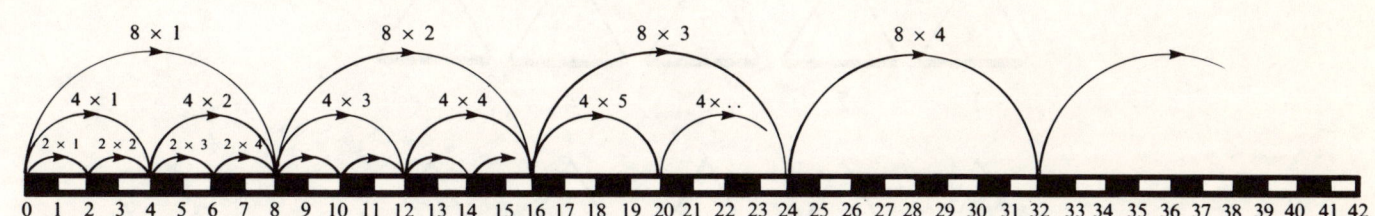

Copy the number line. Continue each table up to 10 times. Make your number line as long as you need.

Fill in the boxes to complete these products. Each set is one product.

1. $8 \times 1 = \mathbf{8}$ **2.** $8 \times 2 = \square$ **3.** $8 \times 3 = \square$ **4.** $8 \times 4 = \square$

 $4 \times 2 = \square$ $4 \times \square = 16$ $4 \times \square = \square$ $4 \times \square = \square$

 $2 \times 4 = \square$ $2 \times \square = \square$ $2 \times 12 = \square$ $2 \times 16 = \square$

G   Continue these sequences.

1. 2, 4, 6, ... 20

2. 4, 8, 12, ... 40

3. 8, 16, 24, ... 80

H   Work out:

1. $\begin{array}{r} 2 \\ \times\,3 \\ \hline \end{array}$ **2.** $\begin{array}{r} 4 \\ \times\,3 \\ \hline \end{array}$ **3.** $\begin{array}{r} 8 \\ \times\,3 \\ \hline \end{array}$ **4.** $\begin{array}{r} 2 \\ \times\,7 \\ \hline \end{array}$ **5.** $\begin{array}{r} 4 \\ \times\,7 \\ \hline \end{array}$

6. $\begin{array}{r} 8 \\ \times\,7 \\ \hline \end{array}$ **7.** $\begin{array}{r} 2 \\ \times\,10 \\ \hline \end{array}$ **8.** $\begin{array}{r} 4 \\ \times\,10 \\ \hline \end{array}$ **9.** $\begin{array}{r} 8 \\ \times\,10 \\ \hline \end{array}$ **10.** $\begin{array}{r} 2 \\ \times\,5 \\ \hline \end{array}$

11. $\begin{array}{r} 4 \\ \times\,5 \\ \hline \end{array}$ **12.** $\begin{array}{r} 8 \\ \times\,5 \\ \hline \end{array}$ **13.** $\begin{array}{r} 2 \\ \times\,1 \\ \hline \end{array}$ **14.** $\begin{array}{r} 4 \\ \times\,1 \\ \hline \end{array}$ **15.** $\begin{array}{r} 8 \\ \times\,1 \\ \hline \end{array}$

I   double 2 → $\square$, double 4 → $\square$, double 8 → $\square$, double $\square$ → 32.

3, 6 and 9 family

A   Look at the triangles. Each has 3 sides.

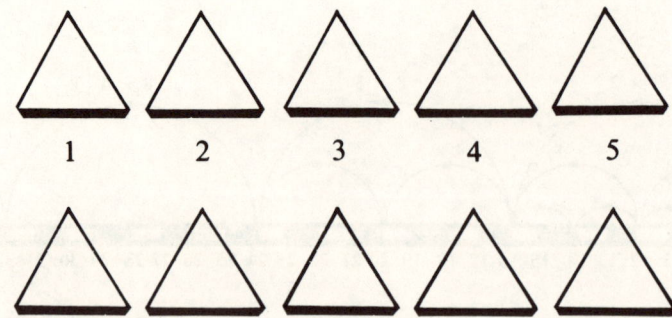

One triangle has 1 × 3 sides ⟶ ☐ sides.
Two triangles have 2 × 3 sides ⟶ ☐ sides.
Three triangles have 3 × 3 sides → ☐ sides.
Four triangles have 4 × 3 sides ⟶ ☐ sides.

Continue this pattern for all 10 triangles.

B   Copy and continue this pattern up to 10 × 3.

|1|1|1| 1 × 3 = ☐
|2| |2| |2| 2 × 3 = ☐
| |3| | |3| | |3| | 3 × 3 = ☐
| |4| | |4| | |4| | 4 × 3 = ☐

C   Complete these:

1. 1 × 3 = ☐   **2.** 5 × 3 = ☐   **3.** 6 × 3 = ☐   **4.** 9 × 3 = ☐
    3 × 1 = ☐   3 × 5 = ☐   3 × 6 = ☐   3 × 9 = ☐
5. 7 × 3 = ☐   **6.** 10 × 3 = ☐   **7.** 0 × 3 = ☐   **8.** 3 × 3 = ☐
    3 × 7 = ☐   3 × 10 = ☐   3 × 0 = ☐

D   Copy and complete:

1. 2 × ☐ = 6   **2.** 3 × ☐ = 12   **3.** ☐ × 3 = 18
4. ☐ × 3 = 9   **5.** ☐ × 3 = 0   **6.** ☐ × 3 = 21
7. ☐ × 3 = 27   **8.** 3 × ☐ = 15   **9.** 3 × ☐ = 24
10. ☐ × 1 = 3   **11.** 3 × ☐ = 30   **12.** 3 × ☐ = 0

E The 9-times table can come from the 10-times table.

Complete the table.

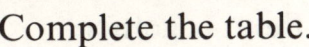
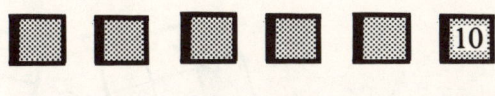

60 50 40 30 20 10

$1 \times 9 = \quad 10 - \quad 1 = \square$

$2 \times 9 = \quad 20 - \quad 2 = \square$
$3 \times 9 = \quad 30 - \quad 3 = \square$
$4 \times 9 = \quad 40 - \quad 4 = \square$
$5 \times 9 = \quad 50 - \quad 5 = \square$
$6 \times 9 = \quad 60 - \quad 6 = \square$
$7 \times 9 = \quad 70 - \quad 7 = \square$
$8 \times 9 = \quad 80 - \quad 8 = \square$
$9 \times 9 = \quad 90 - \quad 9 = \square$
$10 \times 9 = 100 - 10 = \square$

F Copy and complete the mappings.

1.

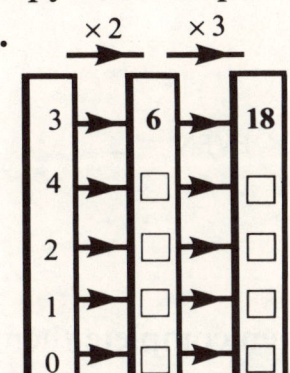

$3 \times 6 = \square$
$4 \times 6 = \square$
$2 \times 6 = \square$
$1 \times 6 = \square$
$0 \times 6 = \square$

2.

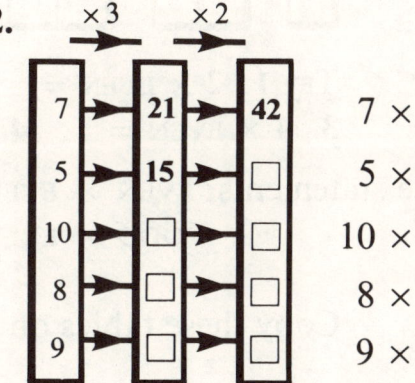

$7 \times 6 = \square$
$5 \times 6 = \square$
$10 \times 6 = \square$
$8 \times 6 = \square$
$9 \times 6 = \square$

G These are number machines.

Put the following numbers into the machines: (a) 3   (b) 7   (c) 8   (d) 9   (e) 0.

Print what comes out.

1.

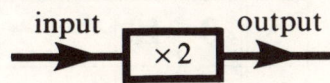

Try:

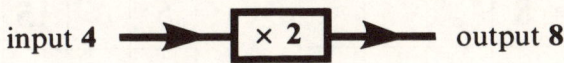

2.

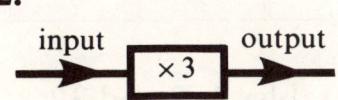

3.

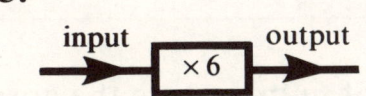

4.

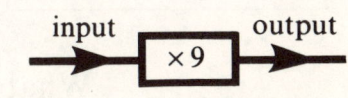

Multiplying patterns

A   Cut out some ODD numbers from centimetre squared paper.
Fit them together.

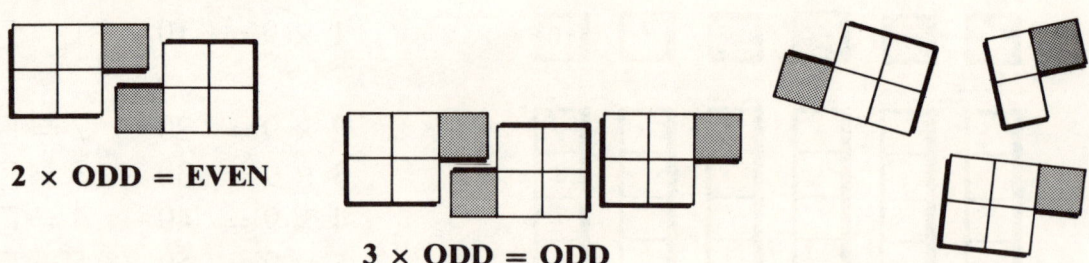

2 × ODD = EVEN

3 × ODD = ODD

Try: **1.** 4 × ODD = __ **2.** 5 × ODD = __ **3.** 6 × ODD = __ **4.** 7 × ODD = __

Complete these statements: ODD × ODD = _____

                                      EVEN × ODD = _____

B Cut out some EVEN numbers from your squared paper.

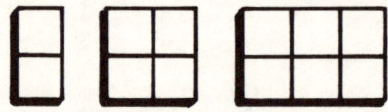

Try **1.** 2 × EVEN = __ **2.** 3 × EVEN = __
3. 4 × EVEN = __ **4.** 5 × EVEN = __ **5.** 6 × EVEN = __

Complete these statements: EVEN × EVEN = _____

                                   ODD × EVEN = _____

C   Copy these tables on your squared paper; then complete them.

1.

×	0	2	4	6	8
0	0	0	0		
2	0				
4	0	8			
6				36	
8					

2.

×	1	3	5	7	9
1					
3				21	
5		15			
7				49	
9					

3.

×	0	2	4	6	8
1	0				
3	0				
5	0	10			
7					56
9					

What kind of numbers are the answers in tables 1, 2 and 3?

D Successive ODD numbers add together to make special
numbers called SQUARE numbers.
Can you see why?
Continue the diagram, and complete the sequence of numbers.

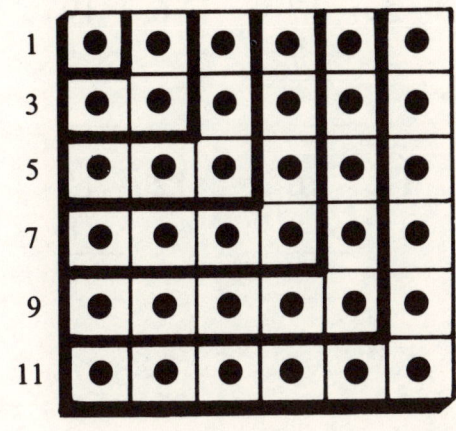

$$1 = \ 1 = \ 1 \times \ 1$$
$$1 + \ 3 = \ 4 = \ 2 \times \ 2$$
$$1 + \ 3 + \ 5 = \ 9 = \ 3 \times \ 3$$
$$1 + \ 3 + \ 5 + \square = 16 = \ 4 \times \ 4$$
$$1 + \ 3 + \ 5 + \square + \square = 25 = \square \times \square$$
$$1 + \ 3 + \ 5 + \square + \square + \square = \square = \square \times \square$$
$$1 + \ 3 + \ 5 + \square + \square + \square + \square = \square = \square \times \square$$
$$1 + \ 3 + \ 5 + \square + \square + \square + \square + \square = \square = \square \times \square$$
$$1 + 3 + \ 5 + \square + \square + \square + \square + \square + \square = \square = \square \times \square$$
$$1 + 3 + 5 + \square + \square + \square + \square + \square + \square + \square = \square = \square \times \square$$

List the set of SQUARE numbers, $\{1, 4, 9, 16, \ldots 100\}$.

E Adding successive EVEN numbers gives a strange pattern.
Complete the table.

$$2 + \ 4 = \ 6 = \ 2 \times \ 3$$
$$2 + \ 4 + \ 6 = 12 = \ 3 \times \ 4$$
$$2 + \ 4 + \ 6 + \ 8 = 20 = \ 4 \times \ 5$$
$$2 + \ 4 + \ 6 + \ 8 + 10 = \square = \ 5 \times \square$$
$$2 + \ 4 + \ 6 + \ 8 + 10 + \square = \square = \square \times \square$$
$$2 + 4 + \ 6 + \ 8 + \square + \square + \square = \square = \square \times \square$$
$$2 + 4 + 6 + \square + \square + \square + \square + \square = \square = \square \times \square$$

F Sometimes *multiplying* is quicker than *adding*.
$$4 + ⑤ + 6 = 5 \times 3 = 15$$
Complete:

1. $1 + ② + 3 = 2 \times \square = \square$ **2.** $8 + ⑨ + 10 = 9 \times \square = \square$

3. $4 + ⑥ + 8 = 6 \times \square = \square$ **4.** $7 + ⑨ + 11 = \square \times \square = \square$

5. $3 + 5 + ⑦ + 9 + 11 = \square \times 5 = \square$ **6.** $2 + 4 + ⑥ + 8 + 10 = \square \times \square = \square$

Multiplying

A *Example:*

$$13 \rightarrow (10 + 3)$$
$$\times\ 4 \qquad \times\ 4$$
$$\overline{} \qquad \overline{40 + 12} \rightarrow (40 + 10 + 2) \rightarrow 52.$$

Work out:

1. 12	**2.** 11	**3.** 15	**4.** 13	**5.** 17	**6.** 19
$\times\ 6$	$\times\ 7$	$\times\ 5$	$\times\ 6$	$\times\ 7$	$\times\ 5$

7. 17	**8.** 13	**9.** 16	**10.** 13	**11.** 16	**12.** 15
$\times\ 6$	$\times\ 7$	$\times\ 7$	$\times\ 8$	$\times\ 8$	$\times\ 8$

B *Example:* $4 \times (2 \times 10) = 4 \times 20 = 80.$

Work out:

1. $2 \times 3 \times 10 = 2 \times 30 = \square$
2. $3 \times 7 \times 10 = 3 \times 70 = \square$
3. $6 \times 7 \times 10 = 6 \times \square = \square$
4. $5 \times 3 \times 10 = 5 \times \square = \square$
5. $3 \times 8 \times 10 = 3 \times \square = \square$
6. $6 \times 5 \times 10 = 6 \times \square = \square$

C Multiply:

1. 20	**2.** 40	**3.** 50	**4.** 30	**5.** 60	**6.** 70
$\times\ 7$	$\times\ 6$	$\times\ 4$	$\times\ 8$	$\times\ 7$	$\times\ 8$

D *Example:*

$$43 \rightarrow (40 + 3)$$
$$\times\ 6 \qquad \times\ 6$$
$$\overline{} \qquad \overline{240 + 18} \rightarrow (240 + 10 + 8) \rightarrow 258.$$

Work out:

1. 24	**2.** 25	**3.** 61	**4.** 72	**5.** 77	**6.** 47
$\times\ 3$	$\times\ 4$	$\times\ 8$	$\times\ 9$	$\times\ 5$	$\times\ 9$

7. 28×3 **8.** 47×5 **9.** 78×9 **10.** 87×8

E Copy and complete the tables.

1. There are 7 days in 1 week.

NO. OF WEEKS	NO. OF DAYS
2	**14**
7	
14	
26	
52	

2. In the dining-room 8 children sit at a table.

NO. OF TABLES USED	NO. OF CHILDREN
2	
7	
9	
12	
14	

3. A gardener plants 16 bulbs in a row.

NO. OF ROWS PLANTED	NO. OF BULBS USED
2	
6	
7	
9	
10	

F (a)

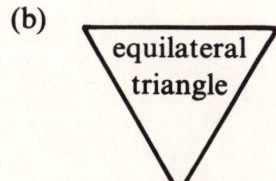

square

(b)

equilateral triangle

Work out the PERIMETER (distance all round) of
(a) a square   (b) an equilateral triangle
with sides:

1. 5 cm long   **2.** 7 cm long   **3.** 9 cm long
4. 14 cm long   **5.** 23 cm long   **6.** 32 cm long.

Polygons

A

Here are five flat shapes.

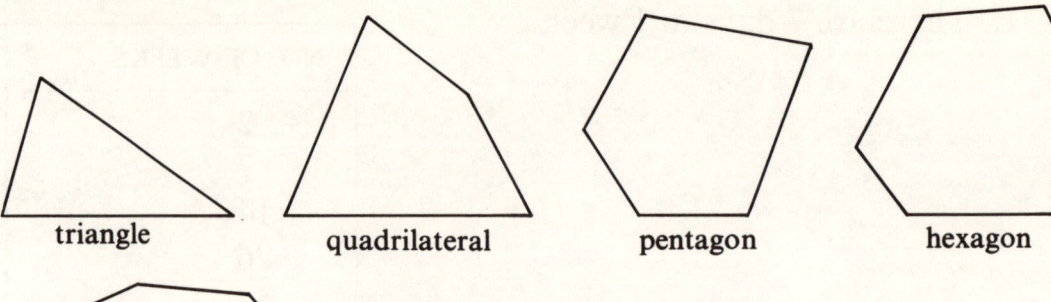

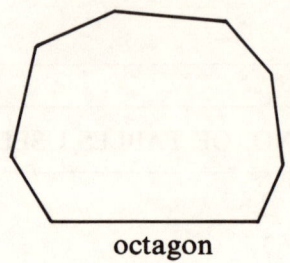

How many sides has:
1. a triangle
2. a quadrilateral
3. a pentagon
4. a hexagon
5. an octagon?

B

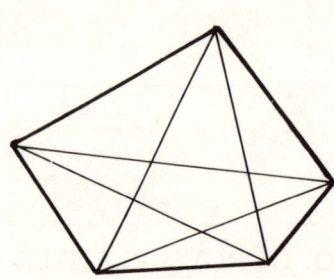

A DIAGONAL is a line that goes from one corner to another.
Copy the five shapes in Section A and draw all the DIAGONALS.

How many DIAGONALS has:

1. a quadrilateral
2. a pentagon
3. a hexagon
4. an octagon?

Why has the triangle been missed out?

C

Copy and complete the table.

SHAPE	NO. OF SIDES	NO. OF ANGLES	NO. OF DIAGONALS
triangle quadrilateral pentagon hexagon octagon			

D All the shapes on the other page are POLYGONS.
When all the sides and all the angles are equal the shapes are
called REGULAR POLYGONS.
Trace and cut out the REGULAR POLYGONS below.

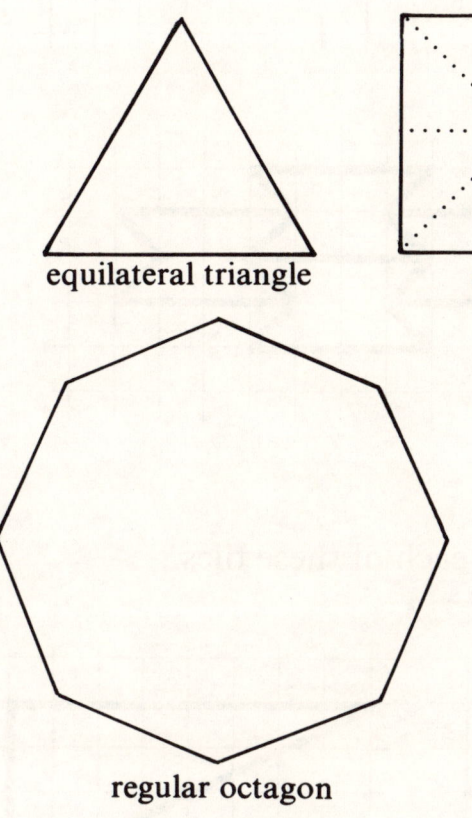

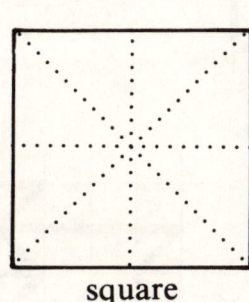

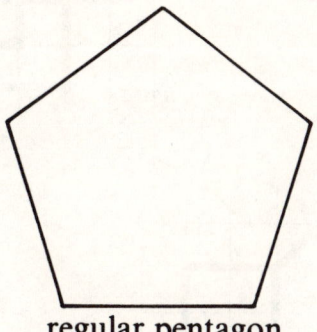

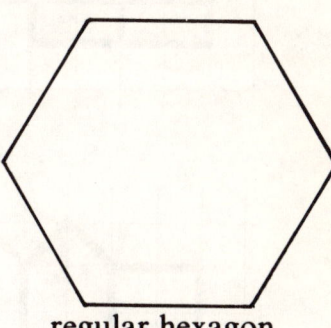

equilateral triangle   square   regular pentagon   regular hexagon

regular octagon

Trace and cut out each POLYGON.
Fold each shape into halves in as many
ways as you can.
The SQUARE shows you – there are 4 ways to
fold it.
How many folds like these can be made on:
1. an equilateral triangle
2. a regular hexagon
3. a regular pentagon
4. a regular octagon?

E Complete the table for REGULAR POLYGONS with these sides:

	PERIMETERS		
LENGTH OF SIDES	regular pentagon	regular hexagon	regular octagon
7 cm			
12 cm			
18 cm			

Tiling patterns

A Copy and continue these patterns. Use squared paper.

1.

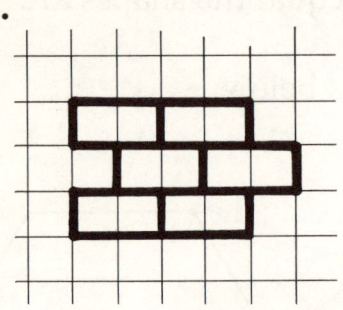

2.

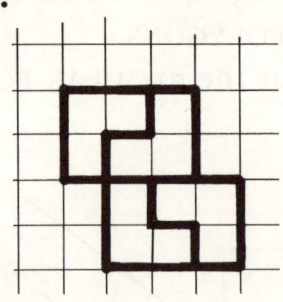

3.

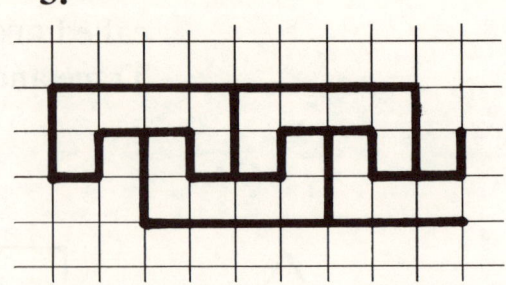

4.

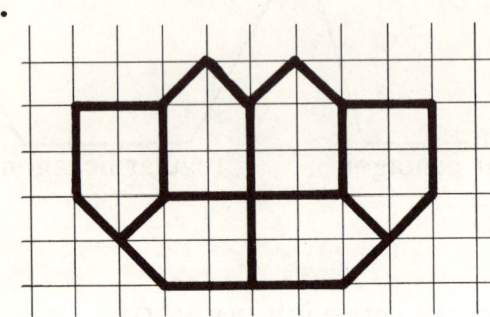

5.

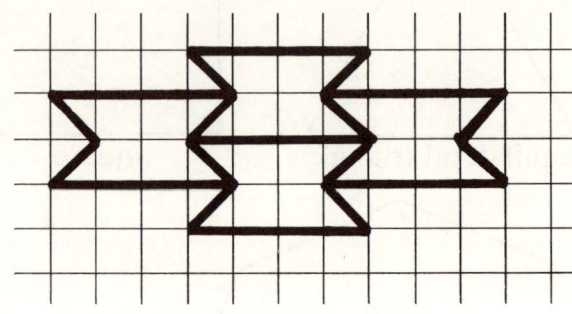

Draw just one of each tile.

B On squared paper make separate tiling patterns from each of these tiles.

1.

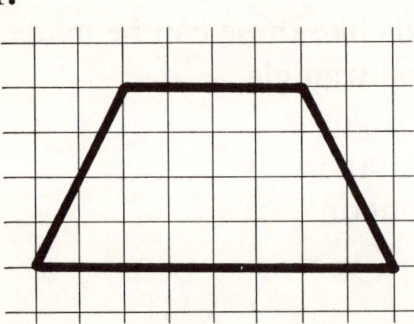

2.

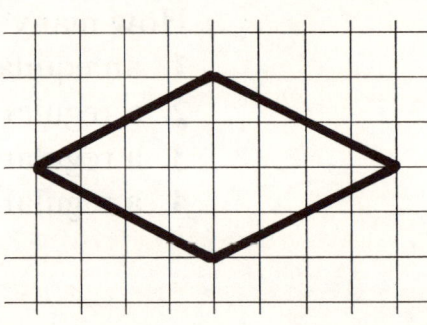

3.

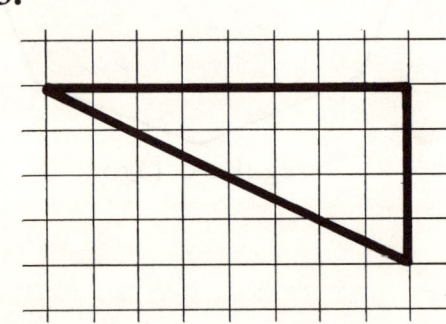

4.

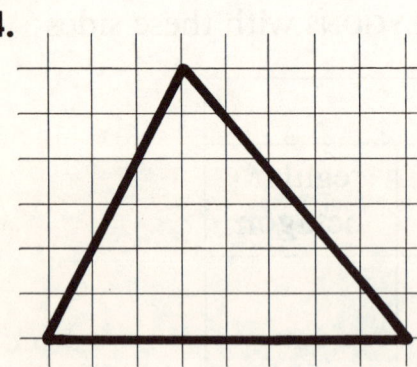

5.

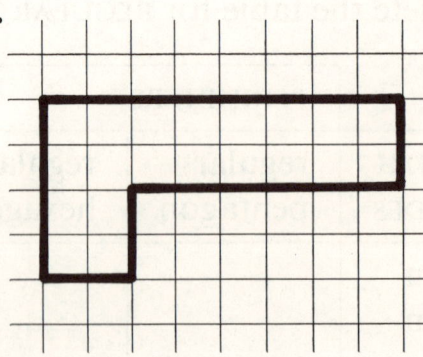

6.

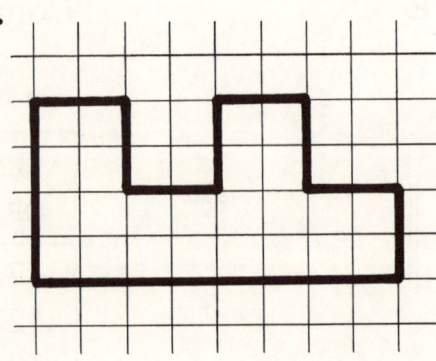

Remember: tiling patterns have no space between the tiles.

C   Trace and cut out a number of these REGULAR POLYGONS.
Try to make tiling patterns with them.

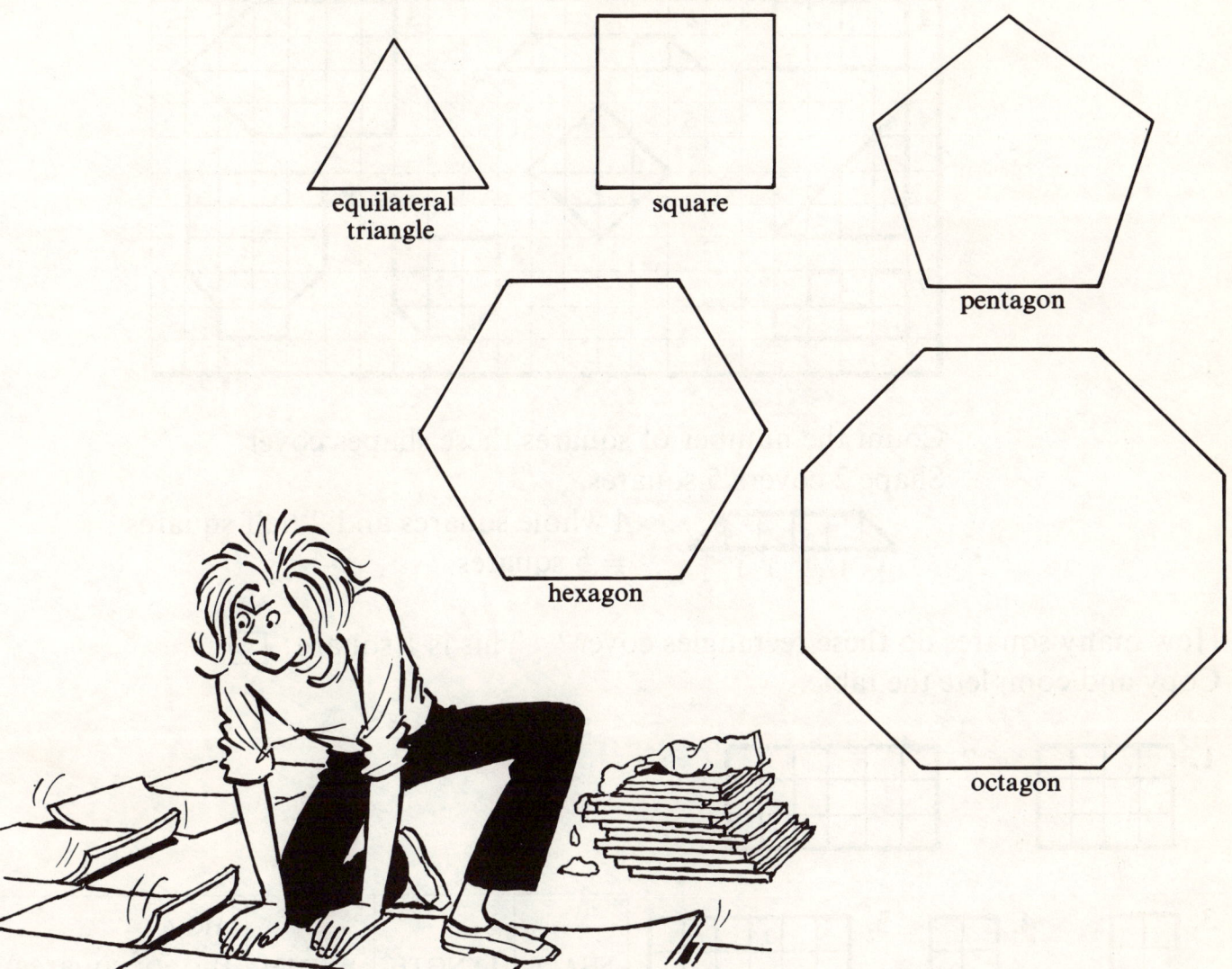

equilateral
triangle

square

pentagon

hexagon

octagon

You can use only three of these shapes for tiling.
Which are they?
Which is the most convenient to use?

D   All of these POLYGONS have sides with the same length.
Try to make tiling patterns using these shapes:

1. octagons and squares
2. hexagons and equilateral triangles
3. hexagons, squares and equilateral triangles.
Can a circle be used for tiling?

Area

A   Shapes are measured by counting the squares they cover.

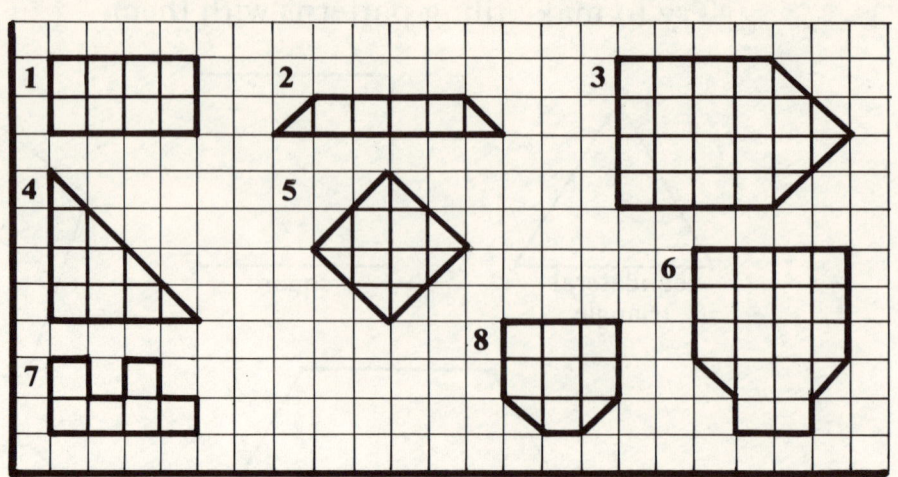

Count the number of squares these shapes cover.

Shape 2 covers 5 squares.

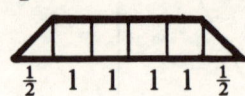

4 whole squares and 2 half squares
= 5 squares

B How many squares do these rectangles cover?   This is 1 square: ☐
Copy and complete the table.

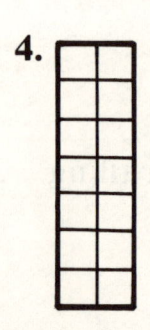
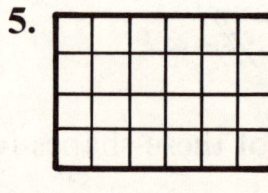

SHAPE	LENGTH	WIDTH	AREA (no. of squares)
1	4	3	12
2			
3			
4			
5			
6			

Complete this sentence:

Area of a RECTANGLE (number of squares) = L_____ × W_____.

C All these shapes cover 6 squares.
Their AREA is 6 unit^2.

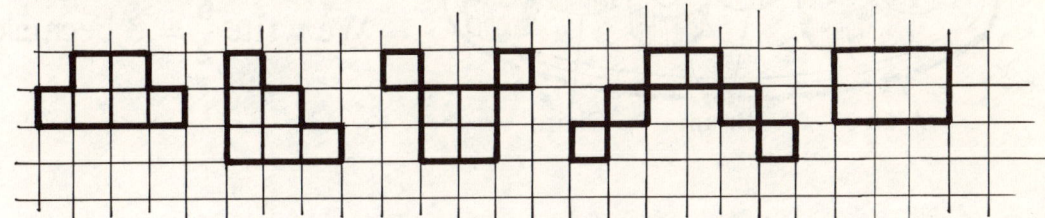

On squared paper draw shapes with an AREA of:

1. 5 unit^2 **2.** 7 unit^2 **3.** 8 unit^2 **4.** 9 unit^2

D Only a rough answer may be given to these shapes.
Find AREA by counting the squares.

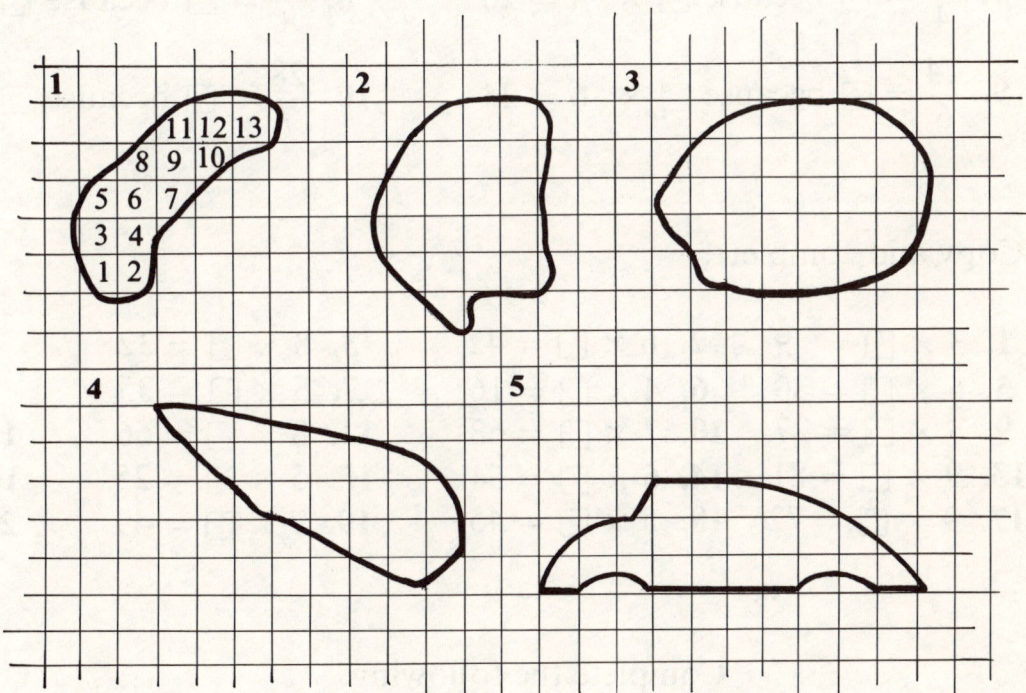

Shape 1 is done for you. Its AREA is about 13 unit^2.
Count as 1 square those bits of squares which are more than half
  a square.
Ignore those bits which are less than half a square.

Dividing

A

Alan   Brian   Claire

6 cakes are shared between Alan, Brian and Claire. They have 2 each.

We write $\dfrac{6}{2} = 3$ because $3 \times 2 = 6$.

Copy and complete.

1. $\dfrac{8}{2} = 4$ because $\square \times 2 = 8$

2. $\dfrac{12}{4} = 3$ because $3 \times \square = \square$

3. $\dfrac{10}{5} = \square$ because $\square \times 5 = \square$

4. $\dfrac{12}{6} = \square$ because $2 \times 6 = \square$

5. $\dfrac{15}{3} = \square$ because $\square \times 3 = 15$

6. $\dfrac{18}{3} = \square$ because $\square \times 3 = \square$

7. $\dfrac{20}{4} = \square$ because $\square \times 4 = 20$

8. $\dfrac{24}{8} = \square$ because $\square \times 8 = 24$

9. $\dfrac{24}{6} = \square$ because $\square \times 6 = 24$

10. $\dfrac{28}{7} = \square$ because $\square \times 7 = 28$

B Copy and complete.

1. $3 \times \square = 9$ 2. $6 \times \square = 42$ 3. $8 \times \square = 32$ 4. $4 \times \square = 36$

5. $6 \times \square = 30$ 6. $4 \times \square = 16$ 7. $5 \times \square = 35$ 8. $8 \times \square = 16$

9. $3 \times \square = 27$ 10. $7 \times \square = 63$ 11. $8 \times \square = 56$ 12. $6 \times \square = 48$

13. $9 \times \square = 81$ 14. $6 \times \square = 54$ 15. $5 \times \square = 25$ 16. $6 \times \square = 36$

17. $9 \times \square = 72$ 18. $5 \times \square = 45$ 19. $7 \times \square = 49$ 20. $7 \times \square = 28$

C Complete the following.

1. $\dfrac{10}{2} = \square$ 2. $\dfrac{14}{2} = \square$ 3. $\dfrac{21}{7} = \square$ 4. $\dfrac{42}{7} = \square$

5. $\dfrac{27}{9} = \square$ 6. $\dfrac{20}{5} = \square$ 7. $\dfrac{56}{7} = \square$ 8. $\dfrac{72}{9} = \square$

9. $\dfrac{24}{3} = \square$ 10. $\dfrac{36}{4} = \square$ 11. $\dfrac{28}{7} = \square$ 12. $\dfrac{42}{6} = \square$

D

This sign ÷ also tells us to share or divide.
Copy and complete these mappings.

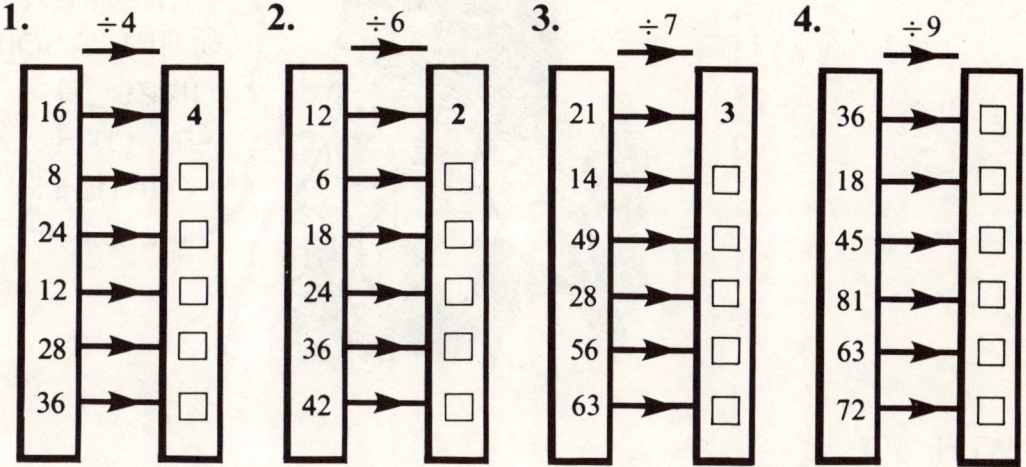

E

There are 24 children in Class I.
How many sets will there be if
their teacher puts 6 children in
each set?

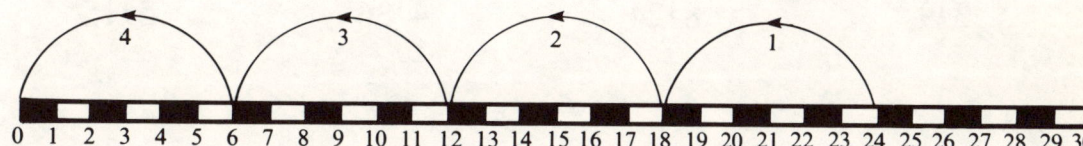

This is written as $\frac{24}{6} = 4$ **sets** because $6 \times 4 = 24$.

Draw similar number lines to show how Class I can be divided
with:

1. 4 in a set   **2.** 3 in a set   **3.** 8 in a set   **4.** 2 in a set.
Record your answers as for the example above.

F

In Class II there are 36 children.
Divide the class up with:
1. 4 in a set   **2.** 6 in a set   **3.** 9 in a set   **4.** 3 in a set.
Use a number line if you wish, but record your answers as in
Section E.

G

Put 7 in a set and find the number of sets for classes of:
1. 21 children   **2.** 35 children   **3.** 28 children   **4.** 42 children.

Practising division

A Example: $24 \div 6$

Write your sum as

$$6\overline{)24}$$
$$\begin{array}{r} 4 \\ 6\overline{)24} \end{array}$$
$6 \times 4 \quad \begin{array}{r} -24 \\ \hline 0 \end{array}$

I must ask myself 6 times something gives 24. Ah! 6 x 4 = 24. I can do it!

Work out:

1. $6\overline{)42}$

2. $5\overline{)20}$

3. $8\overline{)32}$

4. $5\overline{)45}$

5. $4\overline{)16}$

6. $3\overline{)9}$

7. $8\overline{)24}$

8. $4\overline{)36}$

9. $5\overline{)15}$

10. $4\overline{)20}$

11. $6\overline{)18}$

12. $3\overline{)21}$

13. $7\overline{)28}$

14. $6\overline{)54}$

15. $7\overline{)49}$

16. $8\overline{)64}$

17. $6\overline{)36}$

18. $9\overline{)18}$

19. $7\overline{)56}$

20. $9\overline{)63}$

21. $9\overline{)81}$

22. $8\overline{)72}$

23. $10\overline{)40}$

24. $6\overline{)12}$

25. $9\overline{)27}$

B Set down as division sums and work out:

1. $18 \div 2$ **2.** $30 \div 10$ **3.** $54 \div 9$ **4.** $25 \div 5$ **5.** $28 \div 4$

C

1. Work out the number of runs per wicket for each bowler.

BOWLER	RUNS SCORED	WICKETS TAKEN	AVERAGE RUNS PER WICKET
Bumper Bowes	63	7	**9**
Tweaker Smith	42	6	
Basher Barnes	54	9	
Tosser Wicks	72	8	
Lobber Lunn	48	6	

Who had the best bowling average?

2. Find the average number of points scored by these rugby teams.

TEAM	POINTS SCORED	GAMES PLAYED	AVERAGE PER GAME
Leeds	36	6	
St Helen's	56	7	
Widnes	90	10	
Wigan	60	6	
Hull	45	9	

Which team had the best average?

D

There are 28 dominoes in a set.
How many does each player pick up if there are
1. 4 players   2. 7 players   3. 2 players?

E

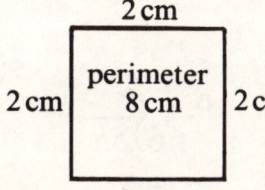

Find the length of the side of each of these squares.
Their PERIMETERS (distance all round) are:
1. 12 cm   2. 20 cm   3. 4 cm   4. 24 cm
5. 32 cm   6. 28 cm   7. 36 cm   8. 40 cm.

Dividing – remainders

A   Divide this 14 cm strip into 3 cm lengths.

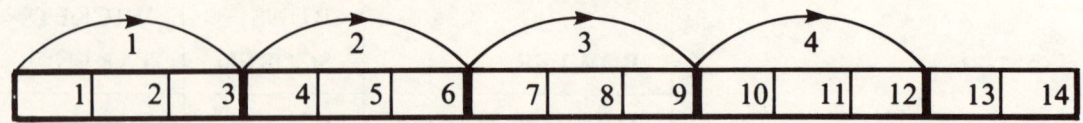

There are 4 pieces 3 cm long, and 2 cm left over.

We write

$$\begin{array}{r} 4 \\ 3\overline{)14} \end{array}$$

$4 \times 3 \quad\quad \dfrac{-12}{2}$ **4 rem 2**

Draw strips of these lengths.
Divide them into 4 cm pieces.
Write the number of 4 cm pieces and the remainder.

1. 10 cm   **2.** 17 cm   **3.** 13 cm   **4.** 21 cm   **5.** 23 cm

Show how you could work out each one *without* drawing.

B Work out:

   1. $17 \div 2 = \mathbf{8}$ rem $\mathbf{1}$ **2.** $21 \div 2 = \square$ rem $\square$ **3.** $14 \div 3 = \square$ rem $\square$

   4. $19 \div 6 = \square$ rem $\square$ **5.** $15 \div 4 = \square$ rem $\square$ **6.** $9 \div 5 = \square$ rem $\square$

   7. $17 \div 7 = \square$ rem $\square$ **8.** $25 \div 6 = \square$ rem $\square$ **9.** $23 \div 5 = \square$ rem $\square$

10. $15 \div 6 = \square$ rem $\square$ **11.** $19 \div 9 = \square$ rem $\square$ **12.** $27 \div 6 = \square$ rem $\square$

13. $25 \div 8 = \square$ rem $\square$ **14.** $22 \div 7 = \square$ rem $\square$ **15.** $22 \div 4 = \square$ rem $\square$

16. $11 \div 3 = \square$ rem $\square$ **17.** $16 \div 5 = \square$ rem $\square$ **18.** $13 \div 6 = \square$ rem $\square$

C Work out these as in Section A.

  1. $8\overline{)35}$ **2.** $6\overline{)29}$ **3.** $7\overline{)36}$ **4.** $7\overline{)39}$ **5.** $5\overline{)42}$

  6. $6\overline{)37}$ **7.** $8\overline{)42}$ **8.** $7\overline{)50}$ **9.** $8\overline{)52}$ **10.** $6\overline{)55}$

D A mechanic had 23 tyres.

He put them on the wheels of 5 cars, 4 wheels per car.

1. How many tyres did he use?

2. How many did he have left over?

E Copy and complete the tables.

1.

NO. OF TYRES	NO. OF CARS	TYRES LEFT OVER
33	8	1
42	9	
25	6	
18	4	
29	7	

2.

NO. OF TYRES	NO. OF 4-WHEELED CARS THAT COULD BE FITTED	TYRES LEFT OVER
17	4	1
27		
35		
37		
41		

F There are 7 days in 1 week.

35 days make 5 weeks $\longrightarrow$ $35 \div 7 = 5$,

36 days make 5 weeks and 1 day $\rightarrow$ $36 \div 7 = 5$ rem 1.

Complete this table.

DAYS	WEEKS and DAYS
22	
30	
50	
60	
65	

Doubling and halving

A   DOUBLING means multiply by 2   ($\times 2$)
HALVING means divide by 2   ($\div 2$)
Copy and complete the arrow graphs.

1.

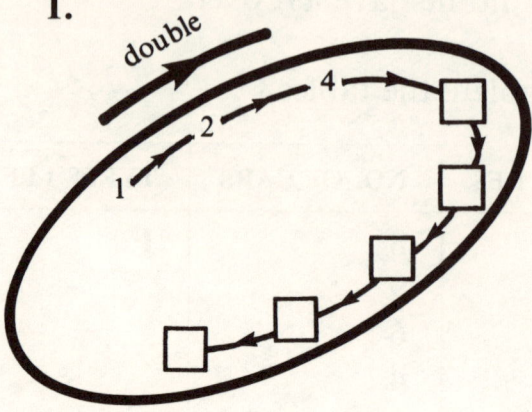

2.

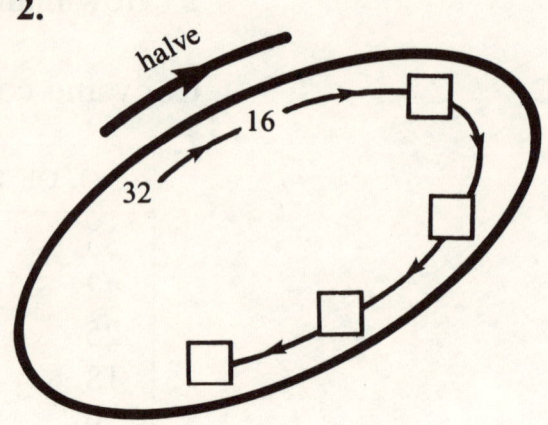

B Measure the lines.   Draw lines DOUBLE the length.

1. _____   ☐ cm
2. _____   ☐ cm
3. _____   ☐ cm
4. _____   ☐ cm
5. _____   ☐ cm

C Measure these lines.   Draw lines HALF the length.

1. _____   ☐ cm
2. _____   ☐ cm
3. _____   ☐ cm
4. _____   ☐ cm
5. _____   ☐ cm

D   1. Double these numbers:
5, 13, 17, 21, 24, 29, 36, 43, 49, 54, 59, 75
2. Halve these:
20, 12, 18, 10, 6, 16, 14, 8, 22, 4, 24, 2

E   Peter's father doubled the money he gave him every day of the week.

He began with 3p.

Copy and complete the table Peter made.

SUNDAY	MONDAY	TUESDAY	WEDNESDAY	THURSDAY	FRIDAY	SATURDAY
3p	p	p	p	p	p	p

How much did he have altogether by   **1.** Tuesday   **2.** Thursday?

F   Sue was given 2 white mice.

They doubled their number every 3 months.

Copy and complete the table.

after	3 months	6 months	9 months	12 months
she had				

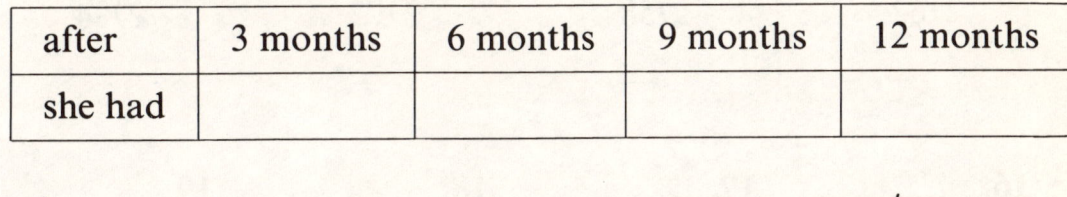

G   Write down the AREA (the number of squares) of the shapes.

Cut out shapes which are   (a) HALF the AREA   (b) DOUBLE the AREA.

1.   **2.** 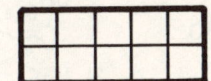   **3.**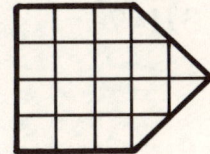

4.    **5.**    **6.**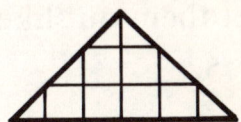

Write the HALVED AREA and DOUBLED AREA in each case.

Dividing and carrying

A *Examples:*   **1.** 34 ÷ 2   **2.** 144 ÷ 6

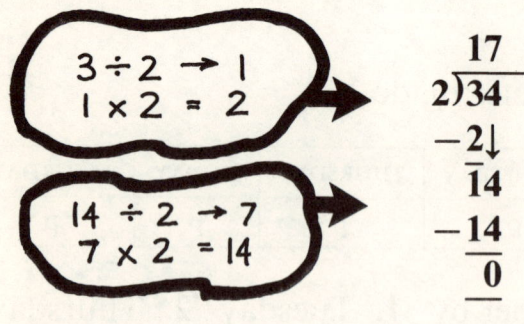

$$\begin{array}{r} 17 \\ 2\overline{)34} \\ -2\downarrow \\ \hline 14 \\ -14 \\ \hline 0 \end{array}$$

$$\begin{array}{r} 24 \\ 6\overline{)144} \\ -12\downarrow \\ \hline 24 \\ -24 \\ \hline 0 \end{array}$$

Work out:

1. 38 ÷ 2   **2.** 54 ÷ 2   **3.** 36 ÷ 2   **4.** 56 ÷ 2   **5.** 44 ÷ 2

6. 24 ÷ 2   **7.** 68 ÷ 2   **8.** 62 ÷ 2   **9.** 74 ÷ 2   **10.** 78 ÷ 2

11. $2\overline{)62}$ **12.** $2\overline{)58}$ **13.** $2\overline{)72}$ **14.** $2\overline{)84}$ **15.** $2\overline{)86}$

16. $2\overline{)96}$ **17.** $2\overline{)98}$ **18.** $2\overline{)104}$ **19.** $2\overline{)114}$ **20.** $2\overline{)132}$

21. $3\overline{)108}$ **22.** $3\overline{)117}$ **23.** $3\overline{)156}$ **24.** $3\overline{)174}$ **25.** $3\overline{)162}$

26. $3\overline{)147}$ **27.** $4\overline{)136}$ **28.** $4\overline{)216}$ **29.** $4\overline{)232}$ **30.** $4\overline{)336}$

B These are also division sums.
Set them out like the others.

1. $\dfrac{52}{4}$ **2.** $\dfrac{96}{4}$ **3.** $\dfrac{75}{5}$ **4.** $\dfrac{90}{5}$ **5.** $\dfrac{85}{5}$

6. $\dfrac{100}{5}$ **7.** $\dfrac{125}{5}$ **8.** $\dfrac{174}{6}$ **9.** $\dfrac{156}{6}$ **10.** $\dfrac{216}{6}$

C *Example:* 197 ÷ 6

```
      32
  6)197
    -18
     17
    -12
      5
```

There are two ways to write the answer:

(a) 32 remainder 5

(b) $32\frac{5}{6}$ (This way tells us to try to divide the 5 into 6 parts.)

Work out and write your answers in both ways:

1. 76 ÷ 6
2. 94 ÷ 6
3. 85 ÷ 7
4. 74 ÷ 4
5. 146 ÷ 5
6. 174 ÷ 7
7. 148 ÷ 9
8. 152 ÷ 7

9. 8)253
10. 9)218
11. 5)329
12. 4)333

13. 3)284
14. 7)222
15. 6)254
16. 9)327

17. $\frac{286}{7}$
18. $\frac{462}{9}$
19. $\frac{324}{5}$
20. $\frac{428}{6}$

21. $\frac{312}{10}$
22. $\frac{464}{10}$
23. $\frac{258}{10}$
24. $\frac{641}{10}$

What do you notice about the answers to the last four?

D

Cedars School has 376 pupils. 9 full coaches will take the children who are going on a trip.

1. How many must each coach carry to take the largest number of children?
2. How many children fill the coaches?
3. How many children do not go?

Parts of a whole

A Copy and complete the chart for these boxing matches.

NUMBER OF ROUNDS	ROUNDS COMPLETED	FRACTION OF FIGHT OVER
2	1	$\frac{1}{2}$
4	1	
3	1	
4	3	
3	2	
5	1	
5	2	
8	1	

B Write in figures:

1. one half 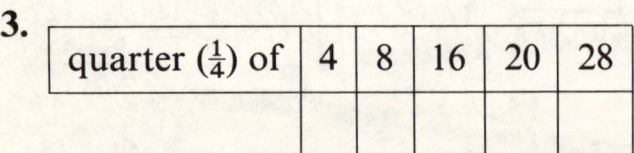 $\frac{1}{2}$ 2. one quarter
3. one third   4. three quarters
5. one eighth   6. two thirds
7. three eighths   8. one tenth
9. seven tenths   10. one sixth

C Copy and complete the tables.

1.

half ($\frac{1}{2}$) of	2	4	8	16	32
	1				

2.

third ($\frac{1}{3}$) of	3	6	9	12	15
	1				

3.

quarter ($\frac{1}{4}$) of	4	8	16	20	28

4.

tenth ($\frac{1}{10}$) of	10	30	50	70	100

5.

eighth ($\frac{1}{8}$) of	8	16	32	48	64

6.

fifth ($\frac{1}{5}$) of	5	20	25	35	40

D

five 2p coins = 10p   2p is $\frac{1}{5}$ of 10p
two 2p coins = 4p   4p is $\frac{2}{5}$ of 10p

1. How many 1p coins make 10p?
What fraction of 10p is (a) 1p (b) 3p (c) 7p (d) 9p?

2. How many 5p coins make 20p?
What fraction of 20p is (a) 5p (b) 10p (c) 15p?

3. How many 10p coins make 50p?
What fraction of 50p is (a) 10p (b) 20p (c) 30p (d) 40p?

4. How many 5p coins make 50p?
What fraction of 50p is (a) 5p (b) 15p (c) 25p (d) 45p?

E   Look at your ruler.

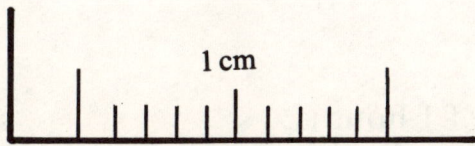

1. How many MILLIMETRES are there in 1 centimetre (cm)?

2. What fraction of 1 centimetre are these:
(a) 1 millimetre (mm) (b) 2 mm (c) 3 mm (d) 5 mm
(e) 7 mm (f) 9 mm?

F   There are 24 hours in a day.
What fraction of a day is:

1. 6 hours   **2.** 3 hours   **3.** 4 hours   **4.** 12 hours
5. 1 hour   **6.** 2 hours   **7.** 8 hours?

Fractions

A

Example: Find $\frac{1}{4}$ of 20p.

20p is the whole.

$\frac{1}{4}$ says 1 ÷ 4 or the whole divided into 4 parts.

$\frac{1}{4}$ of 20p is 20p ÷ 4 = 5p.

Find:

1. $\frac{1}{2}$ of 10p
2. $\frac{1}{4}$ of 12p
3. $\frac{1}{2}$ of 16p
4. $\frac{1}{3}$ of 30p
5. $\frac{1}{3}$ of 24p
6. $\frac{1}{5}$ of 10p
7. $\frac{1}{5}$ of 25p
8. $\frac{1}{10}$ of 50p
9. $\frac{1}{10}$ of 10p
10. $\frac{1}{8}$ of 32p
11. $\frac{1}{6}$ of 36p
12. $\frac{1}{6}$ of 48p.

B

Find the value of:

1. $\frac{1}{5}$ of 20p; $\frac{2}{5}$ of 20p
2. $\frac{1}{3}$ of 15p; $\frac{2}{3}$ of 15p
3. $\frac{1}{8}$ of 16p; $\frac{3}{8}$ of 16p
4. $\frac{1}{4}$ of 40p; $\frac{3}{4}$ of 40p
5. $\frac{1}{10}$ of 50p; $\frac{3}{10}$ of 50p
6. $\frac{1}{6}$ of 24p; $\frac{5}{6}$ of 24p

C

What fraction of a whole turn has the minute hand made turning from:

1.   2.   3.   4.   5.

  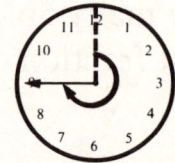  

12 to 6   12 to 3   12 to 9   12 to 4   12 to 8

D

What fraction of 1 hour is:

1. 30 minutes
2. 15 minutes
3. 45 minutes
4. 20 minutes
5. 40 minutes
6. 10 minutes?

E

Copy this 12 cm line.

_____

1. Divide it into: (a) halves (b) quarters (c) thirds
   (d) sixths.
2. How long is: (a) a half (b) a quarter (c) a third
   (d) a sixth (e) two thirds (f) five sixths of the line?

F 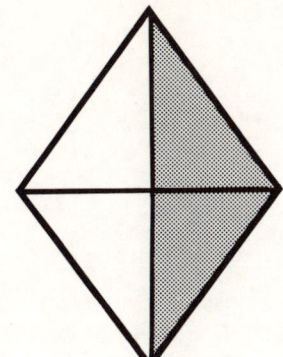 The RHOMBUS is divided into 4 parts.
$\frac{2}{4}$ of the shape is shaded.
$\frac{1}{2}$ of the shape is shaded.
$\frac{1}{2}$ is equivalent to $\frac{2}{4}$.

1. How many parts is each of these shapes divided into?

   (a)   (b)   (c)   (d)   (e)

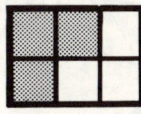 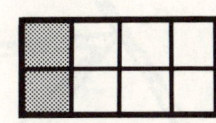  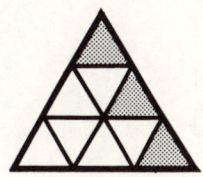 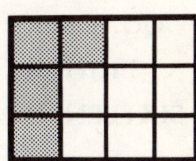

no. of parts ☐   no. of parts ☐   no. of parts ☐   no. of parts ☐   no. of parts ☐

2. Write in 2 ways the fraction of each shape that is *shaded*.

  (a) $\frac{3}{6} = \frac{1}{\square}$ (b) $\frac{2}{\square} = \frac{1}{\square}$ (c) $\frac{2}{\square} = \frac{1}{\square}$ (d) $\frac{3}{\square} = \frac{1}{\square}$ (e) $\frac{4}{\square} = \frac{1}{\square}$

3. Write in 2 ways the fraction that is *not shaded*.

  (a) $\frac{3}{\square} = \frac{1}{\square}$ (b) $\frac{6}{\square} = \frac{3}{\square}$ (c) $\frac{4}{\square} = \frac{2}{3}$ (d) $\frac{6}{\square} = \frac{2}{\square}$ (e) $\frac{8}{\square} = \frac{2}{\square}$

G Copy these letters. Shade the fraction of the AREA given below each.

  1.   **2.**   **3.**   **4.**   **5.**

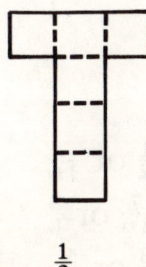   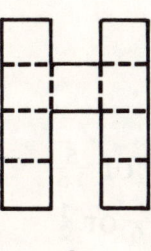      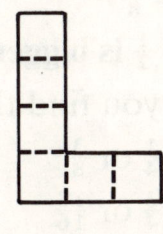   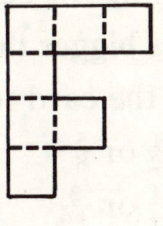

  $\frac{1}{3}$ $\frac{2}{3}$ $\frac{1}{2}$ $\frac{2}{3}$ $\frac{3}{7}$

Equivalent fractions – fraction card

A

ONE WHOLE 1							
HALF				$\frac{1}{2}$			
QUARTER		$\frac{1}{4}$		QUARTER		$\frac{1}{4}$	
$\frac{1}{8}$	$\frac{1}{8}$	$\frac{1}{8}$	$\frac{1}{8}$	$\frac{1}{8}$	$\frac{1}{8}$	$\frac{1}{8}$	$\frac{1}{8}$
$\frac{1}{16}$ $\frac{1}{16}$	$\frac{1}{16}$ $\frac{1}{16}$	$\frac{1}{16}$ $\frac{1}{16}$	$\frac{1}{16}$ $\frac{1}{16}$	$\frac{1}{16}$ $\frac{1}{16}$	$\frac{1}{16}$ $\frac{1}{16}$	$\frac{1}{16}$ $\frac{1}{16}$	$\frac{1}{16}$ $\frac{1}{16}$

Complete the following :
number of parts in a WHOLE

HALVES → ☐
QUARTERS → ☐
EIGHTHS → ☐
SIXTEENTHS → ☐

B Use the Fraction Card to complete :

1. $\frac{1}{2} = \frac{2}{4}$
2. $\frac{1}{4} = \frac{\square}{8}$
3. $\frac{1}{8} = \frac{\square}{16}$
4. $\frac{1}{2} = \frac{\square}{8}$
5. $\frac{1}{4} = \frac{\square}{16}$

6. $\frac{3}{4} = \frac{\square}{8}$
7. $\frac{3}{8} = \frac{\square}{16}$
8. $\frac{5}{8} = \frac{\square}{16}$
9. $\frac{7}{8} = \frac{\square}{16}$
10. $\frac{3}{4} = \frac{\square}{16}$

11. $\frac{1}{2} = \frac{\square}{16}$
12. $1 = \frac{\square}{2}$
13. $1 = \frac{\square}{4}$
14. $1 = \frac{\square}{8}$
15. $1 = \frac{\square}{16}$

C *Example :* Which is larger, $\frac{1}{2}$ or $\frac{3}{8}$?

Using the card, $\frac{1}{2} = \frac{4}{8}$

$\frac{4}{8}$ is bigger than $\frac{3}{8}$ → $\frac{1}{2}$ is bigger than $\frac{3}{8}$

Use the card to help you find the larger of these :

1. $\frac{1}{2}$ or $\frac{5}{8}$
2. $\frac{1}{4}$ or $\frac{3}{8}$
3. $\frac{1}{4}$ or $\frac{1}{8}$
4. $\frac{1}{4}$ or $\frac{5}{16}$
5. $\frac{1}{2}$ or $\frac{7}{16}$

6. $\frac{1}{4}$ or $\frac{3}{16}$
7. $\frac{3}{4}$ or $\frac{13}{16}$
8. $\frac{5}{8}$ or $\frac{9}{16}$
9. $\frac{9}{16}$ or $\frac{1}{2}$
10. $\frac{7}{16}$ or $\frac{3}{8}$

11. $\frac{3}{4}$ or $\frac{9}{16}$
12. $\frac{5}{8}$ or $\frac{7}{16}$
13. $\frac{3}{8}$ or $\frac{5}{16}$
14. $\frac{7}{8}$ or $\frac{3}{4}$
15. $\frac{3}{4}$ or $\frac{5}{8}$

D

ONE WHOLE 1											
HALF						$\frac{1}{2}$					
THIRD			$\frac{1}{3}$		THIRD						
QUARTER		$\frac{1}{4}$		QUARTER			$\frac{1}{4}$				
SIXTH	$\frac{1}{6}$		SIXTH		$\frac{1}{6}$		SIXTH		$\frac{1}{6}$		
$\frac{1}{8}$		$\frac{1}{8}$		$\frac{1}{8}$		$\frac{1}{8}$		$\frac{1}{8}$		$\frac{1}{8}$	
$\frac{1}{12}$	$\frac{1}{12}$	$\frac{1}{12}$	$\frac{1}{12}$	$\frac{1}{12}$	$\frac{1}{12}$	$\frac{1}{12}$	$\frac{1}{12}$	$\frac{1}{12}$	$\frac{1}{12}$	$\frac{1}{12}$	$\frac{1}{12}$

Use the Fraction Card above to fill in the boxes.

1. $\frac{1}{2} = \frac{\square}{6}$ **2.** $\frac{1}{2} = \frac{\square}{12}$ **3.** $\frac{1}{3} = \frac{\square}{6}$ **4.** $\frac{1}{3} = \frac{\square}{12}$ **5.** $\frac{1}{4} = \frac{\square}{12}$

6. $1 = \frac{\square}{3}$ **7.** $1 = \frac{\square}{6}$ **8.** $1 = \frac{\square}{12}$ **9.** $\frac{2}{3} = \frac{\square}{6}$ **10.** $\frac{1}{6} = \frac{\square}{12}$

11. $\frac{5}{6} = \frac{\square}{12}$ **12.** $\frac{3}{4} = \frac{\square}{12}$ **13.** $\frac{2}{3} = \frac{\square}{12}$ **14.** $1 = \frac{\square}{4}$ **15.** $\frac{3}{6} = \frac{\square}{8}$

16. $\frac{3}{6} = \frac{\square}{12}$ **17.** $\frac{4}{8} = \frac{\square}{12}$ **18.** $\frac{2}{4} = \frac{\square}{12}$ **19.** $\frac{2}{8} = \frac{\square}{12}$ **20.** $\frac{6}{8} = \frac{\square}{12}$

E   Fill in the boxes.   You may need both Fraction Cards.

1. $\frac{3}{4} = \frac{\square}{8} = \frac{9}{\square} = \frac{\square}{16}$ **2.** $\frac{1}{3} = \frac{\square}{6} = \frac{4}{\square}$

3. $\frac{1}{2} = \frac{\square}{4} = \frac{3}{\square} = \frac{\square}{8} = \frac{6}{\square} = \frac{\square}{16}$

F   What *one fraction* is EQUIVALENT to:

1. $\frac{2}{4}, \frac{3}{6}, \frac{4}{8}, \frac{6}{12}$ **2.** $\frac{2}{8}, \frac{3}{12}, \frac{4}{16}$ **3.** $\frac{6}{8}, \frac{9}{12}, \frac{12}{16}$?

G $\frac{6}{12}$ is equivalent to $\frac{1}{2}$, $\frac{2}{8}$ is equivalent to $\frac{1}{4}$.
The second fraction is the LOWEST FORM.
Write in its lowest form:

1. $\frac{3}{6}$ **2.** $\frac{2}{6}$ **3.** $\frac{2}{4}$ **4.** $\frac{2}{8}$ **5.** $\frac{2}{12}$ **6.** $\frac{4}{12}$

Equivalent fractions

A Perhaps you have discovered how to find EQUIVALENT FRACTIONS without using your Fraction Cards.

Examples: **1.** $\dfrac{1}{4} = \dfrac{\square}{12}$ **2.** $\dfrac{5}{6} = \dfrac{\square}{12}$

1. $\dfrac{1}{4} = \dfrac{\square}{12}$ ⟨12 is 4×3 I must multiply 1×3⟩ → $\dfrac{1 \times 3}{4 \times 3} = \dfrac{3}{12}$ **2.** $\dfrac{5}{6} = \dfrac{\square}{12}$ ⟨12 is 6×2 I must multiply 5×2⟩ → $\dfrac{5 \times 2}{6 \times 2} = \dfrac{10}{12}$

Complete the following.

1. $\dfrac{1}{4} = \dfrac{\square}{8}$ **2.** $\dfrac{1}{3} = \dfrac{\square}{6}$ **3.** $\dfrac{1}{2} = \dfrac{\square}{10}$ **4.** $\dfrac{1}{5} = \dfrac{\square}{10}$ **5.** $\dfrac{1}{6} = \dfrac{\square}{12}$

6. $\dfrac{1}{2} = \dfrac{\square}{6}$ **7.** $\dfrac{1}{4} = \dfrac{\square}{16}$ **8.** $\dfrac{1}{3} = \dfrac{\square}{12}$ **9.** $\dfrac{1}{2} = \dfrac{\square}{8}$ **10.** $\dfrac{1}{8} = \dfrac{\square}{16}$

11. $\dfrac{2}{3} = \dfrac{\square}{12}$ **12.** $\dfrac{2}{5} = \dfrac{\square}{10}$ **13.** $\dfrac{3}{4} = \dfrac{\square}{16}$ **14.** $\dfrac{3}{4} = \dfrac{\square}{12}$ **15.** $\dfrac{2}{3} = \dfrac{\square}{6}$

16. $\dfrac{3}{4} = \dfrac{\square}{8}$ **17.** $\dfrac{3}{5} = \dfrac{\square}{10}$ **18.** $\dfrac{1}{2} = \dfrac{\square}{16}$ **19.** $\dfrac{4}{5} = \dfrac{\square}{10}$ **20.** $\dfrac{3}{8} = \dfrac{\square}{16}$

B Fractions usually have to be written in their lowest form.

Example: $\dfrac{6}{12} = \dfrac{\square}{\square}$ ⟨What is the highest number I can divide top and bottom by?⟩ → ⟨Aha! ÷6⟩ → $\dfrac{6 \div 6}{12 \div 6} = \dfrac{1}{2}$

Change to their lowest form:

1. $\dfrac{4}{8} = \dfrac{\square}{\square}$ **2.** $\dfrac{2}{6} = \dfrac{\square}{\square}$ **3.** $\dfrac{2}{10} = \dfrac{\square}{\square}$ **4.** $\dfrac{3}{12} = \dfrac{\square}{\square}$ **5.** $\dfrac{4}{12} = \dfrac{\square}{\square}$

6. $\dfrac{5}{10} = \dfrac{\square}{\square}$ **7.** $\dfrac{8}{12} = \dfrac{\square}{\square}$ **8.** $\dfrac{6}{8} = \dfrac{\square}{\square}$ **9.** $\dfrac{4}{6} = \dfrac{\square}{\square}$ **10.** $\dfrac{8}{16} = \dfrac{\square}{\square}$

11. $\dfrac{12}{16} = \dfrac{\square}{\square}$ **12.** $\dfrac{6}{10} = \dfrac{\square}{\square}$ **13.** $\dfrac{4}{10} = \dfrac{\square}{\square}$ **14.** $\dfrac{6}{16} = \dfrac{\square}{\square}$ **15.** $\dfrac{4}{16} = \dfrac{\square}{\square}$

16. $\dfrac{9}{12} = \dfrac{\square}{\square}$ **17.** $\dfrac{4}{4} = \square$ **18.** $\dfrac{5}{5} = \square$ **19.** $\dfrac{8}{8} = \square$ **20.** $\dfrac{12}{12} = \square$

C Copy and complete these sentences.

1. $\frac{1}{4}$ is $\frac{\square}{8}$ and **twice** as large as $\frac{1}{8}$

2. $\frac{1}{2}$ is $\frac{\square}{4}$ and _____ as large as $\frac{1}{4}$

3. $\frac{1}{8}$ is $\frac{\square}{16}$ and _____ as large as $\frac{1}{16}$

4. $\frac{1}{2}$ is $\frac{\square}{8}$ and **four times** as large as $\frac{1}{8}$

5. $\frac{1}{4}$ is $\frac{\square}{16}$ and _____ as large as $\frac{1}{16}$

6. $\frac{1}{2}$ is $\frac{\square}{16}$ and _____ as large as $\frac{1}{16}$

7. $\frac{1}{2}$ is $\frac{\square}{6}$ and _____ as large as $\frac{1}{6}$

8. $\frac{1}{4}$ is $\frac{\square}{12}$ and _____ as large as $\frac{1}{12}$

9. $\frac{1}{3}$ is $\frac{\square}{6}$ and _____ as large as $\frac{1}{6}$

10. 1 is $\frac{\square}{4}$ and _____ as large as $\frac{1}{4}$

D Write one fraction which is EQUIVALENT to each set of fractions.

1. $\{\frac{2}{4}, \frac{3}{6}, \frac{4}{8}, \frac{5}{10}\}$

2. $\{\frac{2}{6}, \frac{3}{9}, \frac{4}{12}, \frac{5}{15}\}$

3. $\{\frac{2}{10}, \frac{3}{15}, \frac{4}{20}, \frac{5}{25}\}$

4. $\{\frac{2}{8}, \frac{3}{12}, \frac{4}{16}, \frac{5}{20}\}$

5. $\{\frac{2}{12}, \frac{3}{18}, \frac{4}{24}, \frac{5}{30}\}$

6. $\{\frac{2}{20}, \frac{3}{30}, \frac{4}{40}, \frac{5}{50}\}$

E Copy and complete:

1. $\frac{1}{12}$ hour is **5 minutes** → $\frac{5}{12}$ hour is $\square$ minutes.

2. $\frac{6}{12}$ hour is $\square$ minutes, which is $\square$ an hour.

3. $\frac{3}{12}$ hour is $\square$ minutes, which is $\square$ an hour.

4. $\frac{7}{12}$ hour is $\square$ minutes.

5. $\frac{11}{12}$ hour is $\square$ minutes.

6. $\frac{9}{12}$ hour is $\square$ minutes, which is $\square$ an hour.

7. $\frac{4}{12}$ hour is $\square$ minutes, which is $\square$ an hour.

8. $\frac{8}{12}$ hour is $\square$ minutes, which is $\square$ an hour.

9. $\frac{10}{12}$ hour is $\square$ minutes, which is $\square$ an hour.

10. $\frac{2}{12}$ hour is $\square$ minutes, which is $\square$ an hour.

Units of length

A

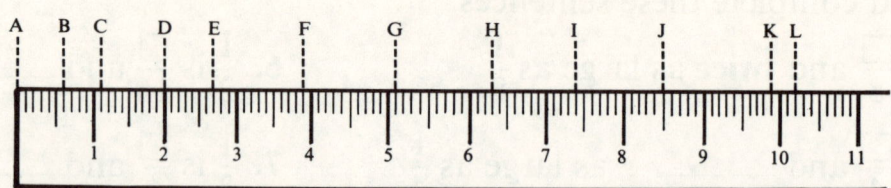

This ruler measures CENTIMETRES (cm) and MILLIMETRES (mm).

10 millimetres (10 mm) = 1 centimetre (1 cm)

1 millimetre   (1 mm) = $\frac{1}{10}$ cm   or   0·1 cm

2 millimetres (2 mm) = $\frac{2}{10}$ cm   or   0·2 cm

Copy and complete the table.

	cm	mm	cm (fraction)	cm (decimal)	mm
A → B	0	6	$\frac{6}{10}$	0·6	6
A → C	1	1	$1\frac{1}{10}$	1·1	
A → D				2·0	
A → E			$2\frac{7}{10}$		
A → F					39
A → G					
A → H					
A → I					
A → J					
A → K				9·9	
A → L					

B

Write as MILLIMETRES (mm):

1. 0·4 cm   **2.** 0·7 cm   **3.** 0·5 cm   **4.** 0·8 cm   **5.** 0·3 cm

C

Write as CENTIMETRES (cm):

1. 6 mm   **2.** 5 mm   **3.** 9 mm   **4.** 8 mm   **5.** 4 mm

D

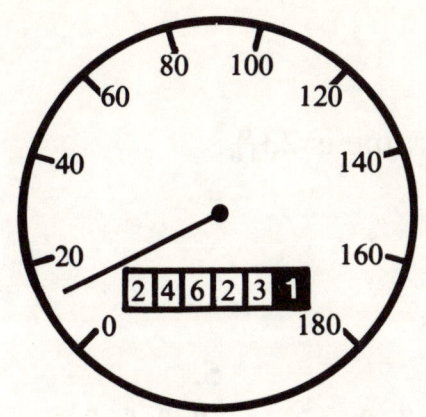

A car's speedometer has a kilometre measurer called an ODOMETER.

ODOMETER

This car has travelled
24623·1 kilometres (km) or $24623\frac{1}{10}$ km
Write in figures in 2 ways what these ODOMETERS read.

1.
| 2 | 1 | 5 | 3 | 6 | 2 |

2.
| 1 | 1 | 0 | 2 | 4 | 7 |

3.
| 0 | 9 | 1 | 2 | 6 | 5 |

4.
| 5 | 8 | 0 | 0 | 2 | 9 |

5.
| 3 | 3 | 0 | 5 | 7 | 6 |

6.
| 1 | 8 | 0 | 0 | 0 | 4 |

7.
| 0 | 0 | 3 | 2 | 4 | 6 |

8.
| 0 | 0 | 0 | 1 | 8 | 0 |

Which car was probably the newest?
Which car was probably the oldest?

E

On squared paper draw ODOMETERS to show these readings. They are in kilometres (km).
Draw the tenths in another colour.

  1. 52041·7   **2.** 84025·3   **3.** 2409·8

  4. 10000·0   **5.** 5109·2   **6.** 117·1

  7. 25·5   **8.** 25010·9   **9.** 67240·2

  10. $32445\frac{6}{10}$ **11.** $15608\frac{4}{10}$ **12.** $2560\frac{8}{10}$

  13. $428\frac{3}{10}$ **14.** $24264\frac{5}{10}$ **15.** $60606\frac{9}{10}$

  16. $12547\frac{1}{10}$ **17.** $6000\frac{2}{10}$ **18.** $17524\frac{7}{10}$

F

Study these examples: **3648·5 km ⟶ $3648\frac{5}{10}$ km ⟶ $3648\frac{1}{2}$ km**

51742·2 km → $51742\frac{2}{10}$ km → $51742\frac{1}{5}$ km

Write these distances in 3 ways as above.

  1. 17503·6 km   **2.** 6214·5 km   **3.** 305·4 km   **4.** 6129·8 km

Decimal fractions – tenths

A

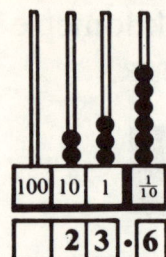

The number shown on the counting frame is $23\frac{6}{10}$.
This is written **23·6**
and said, **'Twenty-three point six.'**

Copy and complete the boxes below each counting frame.

1.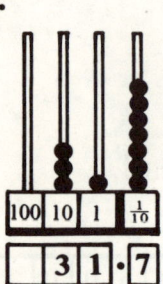

2.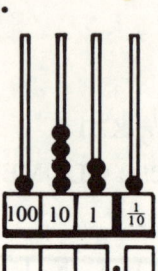

3.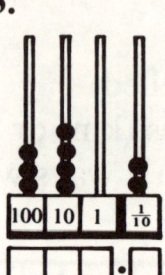

4.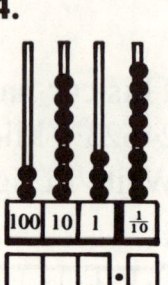

5.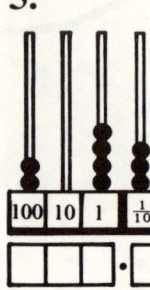

6.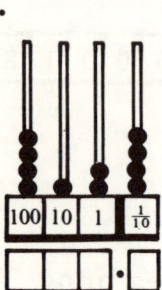

7.

8.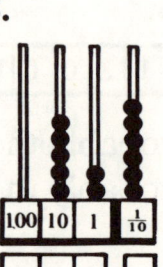

9.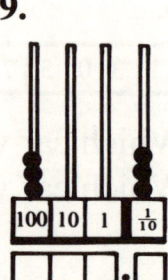

10.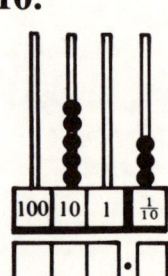

B   Draw counting frames to show:

    1. 14·2   **2.** 114·6   **3.** 50·4   **4.** 7·6

    5. 142·0   **6.** 403·7   **7.** 240·2   **8.** 666·6

C   Draw counting frames to show:

    1. $15\frac{1}{10}$ **2.** $102\frac{3}{10}$

    3. $243\frac{7}{10}$ **4.** $100\frac{9}{10}$.

D   Draw counting frames to show:

    1. $21\frac{1}{5}(\frac{2}{10})$ **2.** $104\frac{3}{5}$

    3. $400\frac{1}{2}$ **4.** $29\frac{4}{5}$.

72

E

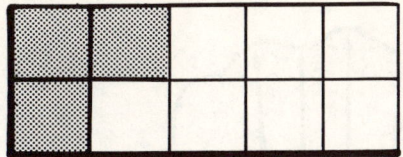

In the large rectangle
3 of the small squares are shaded.
3 out of 10 are shaded → $\frac{3}{10}$ → 0·3.

On squared paper draw rectangles with 10 small squares.
Shade:

1. 4 small squares   **2.** 7 small squares   **3.** 5 small squares
4. 9 small squares   **5.** 6 small squares   **6.** 2 small squares
7. 8 small squares   **8.** 10 small squares   **9.** 1 small square.

Copy and complete the table.
Write the areas you have shaded first as common fractions of the rectangle, and then as decimal fractions of the rectangle.

	NUMBER OF SMALL SQUARES	NUMBER OF $\frac{1}{10}$'s	FRACTION IN LOWEST FORM	DECIMAL FRACTION
1.	4	$\frac{4}{10}$	$\frac{2}{5}$	0·4
2.	7			
3.				
4.				
5.				
6.				
7.				
8.				
9.				

F

0·1 of 10p is $\frac{1}{10}$ **of 10p** → **1p.**
Work out:

1. 0·3 of 10p   **2.** 0·7 of 10p   **3.** 0·1 of 50p   **4.** 0·5 of 50p
5. 0·7 of 50p   **6.** 0·8 of 50p   **7.** 0·5 of 2p   **8.** 0·2 of 5p

The calendar

A

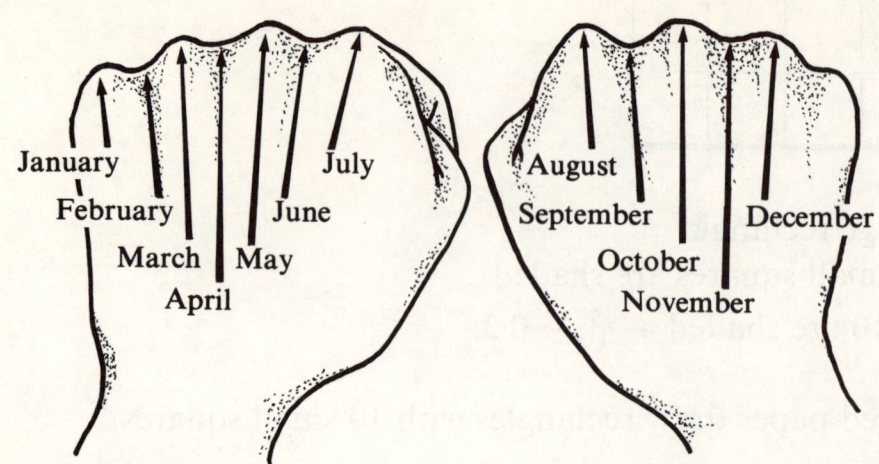

Here is a way to remember the long months and the short months of the year.
Clench your fists.
Look at your knuckles.
The high points are those months with 31 days.

1. List all the months of the year with 31 days.
   {January, ..., ... December}
2. List all the months of the year with 30 days (*not* February).
   {April, ..., ... November}
3. How many days are there altogether in the months with 31 days?
4. How many days are there altogether in the months with 30 days?

B February has 28 days and 29 in a LEAP YEAR.
Leap years occur every 4 years. 1980 was a LEAP YEAR.
1. Write the 4 LEAP YEARS after 1980.
2. Write the 4 LEAP YEARS before 1980.
3. Pick out the LEAP YEARS from this set of years:
   {1976, 1978, 1980, 1982, 1984, 1986, 1988, 1990}
4. Will the year 2000 be a LEAP YEAR?

C Work out how many days there are in these months altogether.
1. June, July, August   2. April, May, June
3. October, November, December 4. January, February, March
5. November, December, January 6. December, January, February (1985).

D School holidays are from 23rd July to 5th September inclusive.
How many days is this?

E

Make sure you know your 7-times table.

Today is Sunday. What day will it be after these times? (The first is done for you.)

1. 15 days (15 ÷ 7 = 2 rem 1) → **2 weeks and 1 day** → **Monday**

2. 7 days	**3.** 9 days	**4.** 11 days	**5.** 14 days
6. 16 days	**7.** 21 days	**8.** 25 days	**9.** 28 days
10. 31 days	**11.** 32 days	**12.** 35 days	**13.** 39 days
14. 40 days	**15.** 42 days	**16.** 45 days	**17.** 49 days
18. 55 days	**19.** 60 days	**20.** 65 days	**21.** 70 days

F

How many days are there from:

1. 28th January to 5th February?

2. 23rd May to 7th June?

3. 17th April to 10th May?

4. 29th July to 2nd September?

5. 25th December to 6th February?

G

1. When 28th January is Monday, what day is 5th February?

2. When 23rd May is Friday, what day is 7th June?

3. When 17th April is Tuesday, what day is 10th May?

4. When 29th July is Sunday, what day is 2nd September?

5. When 25th December is Friday, what day is 6th February?

H

What will the date be:

1. 2 weeks after 31st January?

2. 4 weeks after 13th January?

3. 2 weeks after 21st February 1985?

4. 5 weeks after 24th March?

5. 6 weeks after 22nd July?

6. 2 weeks *before* Christmas Day?

Solid shapes

A

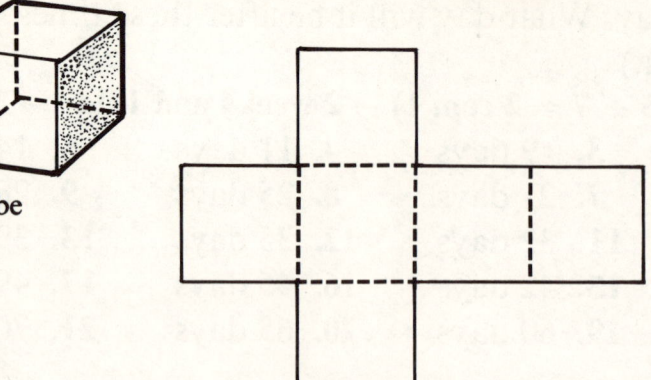

cube

A CUBE is a solid shape.

This is the NET of a CUBE (the CUBE opened out).
The CUBE has
   6 square FACES
   8 VERTICES (corners)
12 EDGES.

1. Think of 3 common articles that are CUBE shaped.
2. On a piece of card draw the NET of the CUBE. Cut it out. Fold it to make a CUBE.

B

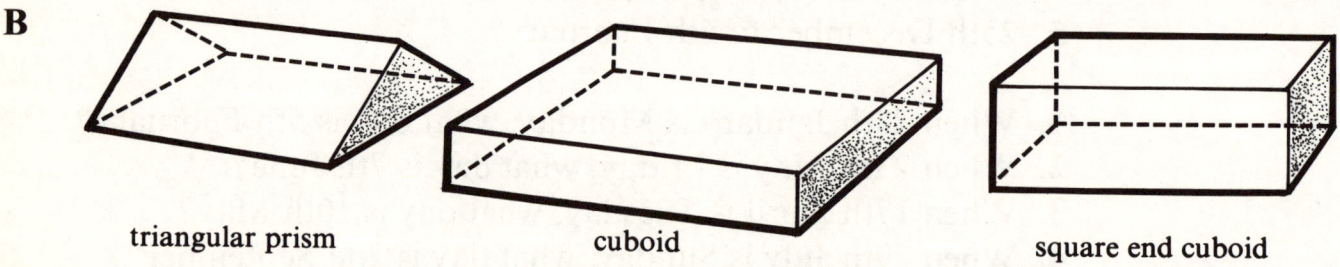

triangular prism   cuboid   square end cuboid

1. Suggest articles that are packed using each of the shapes above.
2. Make freehand sketches of the NET of each shape.
3. How many FACES has each shape?
4. How many VERTICES (corners) has each shape?
5. How many EDGES has each shape?

All these shapes are PRISMS.

C These are nets of PYRAMIDS.
Trace them on to card. Cut out the nets.
Fold the nets to make solid shapes.

1.   **2.**

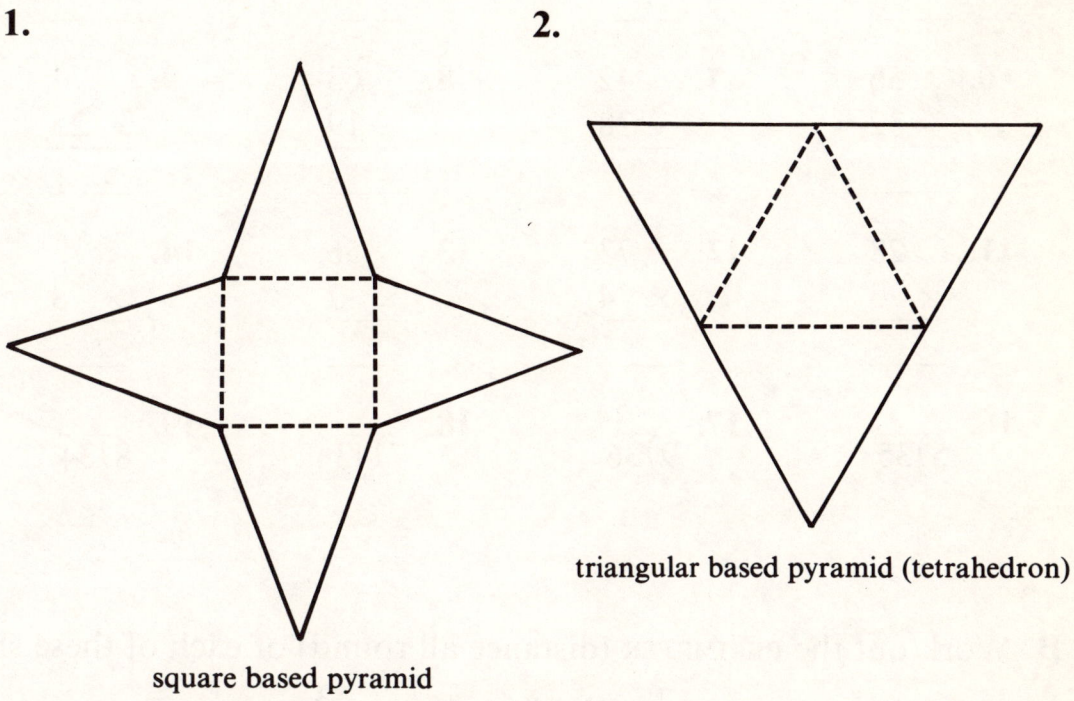

triangular based pyramid (tetrahedron)

square based pyramid

The shapes on the opposite page are all PRISMS.
Can you see how PRISMS differ from PYRAMIDS?

D Copy and complete the table showing which flat shapes make up
the nets.

SHAPES	RECTANGLES	SQUARES	TRIANGLES
cube cuboid square ended cuboid triangular prism square based pyramid triangular based pyramid			

77

Revision

A Work out:

1. 34 + 27	**2.** 29 + 36	**3.** 48 + 77	**4.** 69 + 78	**5.** 94 + 63
6. 56 − 32	**7.** 42 − 28	**8.** 63 − 19	**9.** 74 − 25	**10.** 99 − 66
11. 28 × 3	**12.** 73 × 4	**13.** 56 × 7	**14.** 73 × 8	**15.** 37 × 9
16. 5)35	**17.** 9)36	**18.** 7)43	**19.** 8)34	**20.** 9)65

B Work out the PERIMETER (distance all round) of each of these shapes.

1.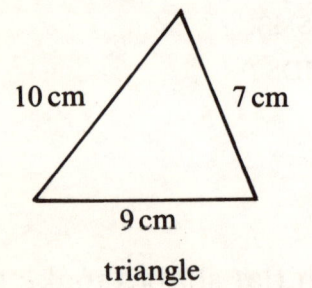

10 cm   7 cm   9 cm

triangle

2.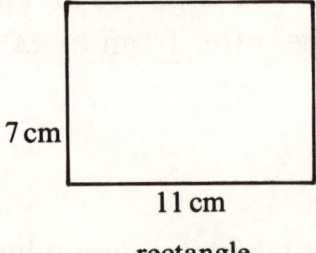

7 cm   11 cm

rectangle

3.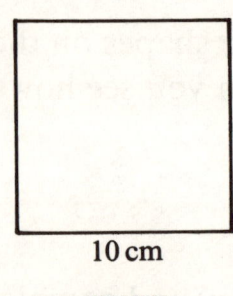

10 cm

square

4.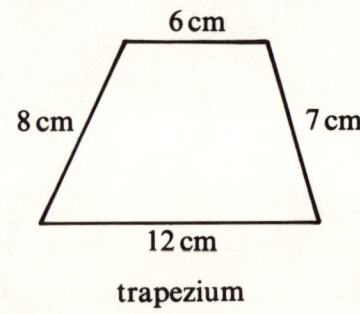

6 cm   8 cm   7 cm   12 cm

trapezium

C First DOUBLE, then HALVE, each of the following.

    1. 12   **2.** 18   **3.** 32   **4.** 42   **5.** 66   **6.** 74

D   **1.** Is 76 nearer to 70 or to 80?   **2.** Is 44 nearer to 40 or to 50?
    3. Is 31 nearer to 30 or to 40?   **4.** Is 58 nearer to 50 or to 60?
    5. Is 63 nearer to 60 or to 70?   **6.** Is 77 nearer to 70 or to 80?

E

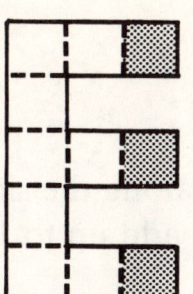 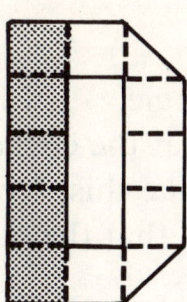 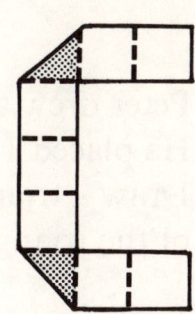 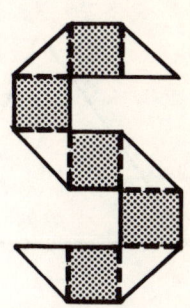

1. How many whole squares does each letter cover?
   (Remember 2 half squares make 1 whole square.)
   Write your answers like this: **E → 11**
2. How many whole squares are shaded in each letter?
3. What fraction of each letter is shaded?
4. What fraction of each letter is unshaded?

F Write the following in their lowest form.

1. $\dfrac{5}{10} = \dfrac{1}{\Box}$ 2. $\dfrac{2}{10} = \dfrac{1}{\Box}$ 3. $\dfrac{2}{6} = \dfrac{1}{\Box}$ 4. $\dfrac{2}{8} = \dfrac{1}{\Box}$ 5. $\dfrac{4}{12} = \dfrac{1}{\Box}$

6. $\dfrac{6}{8} = \dfrac{3}{\Box}$ 7. $\dfrac{4}{10} = \dfrac{2}{\Box}$ 8. $\dfrac{8}{12} = \dfrac{2}{\Box}$ 9. $\dfrac{9}{12} = \dfrac{3}{\Box}$ 10. $\dfrac{8}{10} = \dfrac{4}{\Box}$

G Write the following using decimals.

1. $\dfrac{1}{10} = 0\cdot\Box$ 2. $\dfrac{3}{10} = 0\cdot\Box$ 3. $\dfrac{9}{10} = 0\cdot\Box$ 4. $\dfrac{7}{10} = 0\cdot\Box$

5. $\dfrac{1}{5} = \dfrac{\Box}{10} = 0\cdot\Box$ 6. $\dfrac{4}{5} = \dfrac{\Box}{10} = 0\cdot\Box$ 7. $\dfrac{1}{2} = \dfrac{\Box}{10} = 0\cdot\Box$ 8. $\dfrac{3}{5} = \dfrac{\Box}{10} = 0\cdot\Box$

H Find these fractional parts.

1. $\frac{1}{2}$ of 10p = $\Box$ p 2. $\frac{1}{5}$ of 10p = $\Box$ p 3. $\frac{1}{4}$ of 20p = $\Box$ p
4. $\frac{1}{10}$ of 1 cm = $\Box$ mm 5. $\frac{1}{5}$ of 1 cm = $\Box$ mm 6. $\frac{3}{10}$ of 1 cm = $\Box$ mm
7. $\frac{1}{2}$ of 1 cm = $\Box$ mm 8. $\frac{3}{5}$ of 1 cm = $\Box$ mm 9. $\frac{4}{5}$ of 1 cm = $\Box$ mm
10. 0·1 of 1 cm = $\Box$ mm 11. 0·3 of 1 cm = $\Box$ mm 12. 0·9 of 1 cm = $\Box$ mm

Puzzles

A

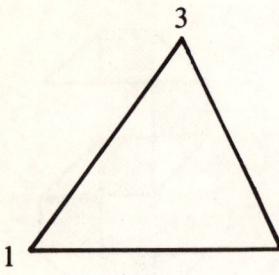

Peter drew a triangle.
He placed 1, 2, 3 at the corners.
Draw a triangle like this. Now place 4, 5, 6, 7, 8, 9 along the sides of the triangle, so that the numbers along each side add up to 17.

B Copy these sums and fill in the missing numbers.

1.
```
   □ 9
 + 3 □
 ─────
   6 4
```

2.
```
   4 □
 - □ 9
 ─────
   2 3
```

3.
```
   2 □
 ×   4
 ─────
 1 □ 8
```

4.
```
      5
 7)□ □
```

C

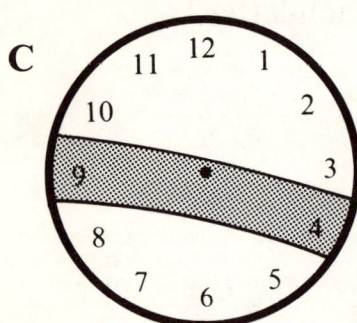

Peter had a clock for his birthday.
Unfortunately he dropped it and the face broke into 6 pieces.
Each piece had 2 numbers adding up to 13.
Copy the clock face and draw lines to show how it was broken.

D Copy and complete the cross-number puzzle.

Clues across
- **2** 72 × 2
- **4** 33 ÷ 3
- **7** 56 + 54
- **9** 47 × 2 × 3
- **11** 8 × 8
- **13** 64 + 36

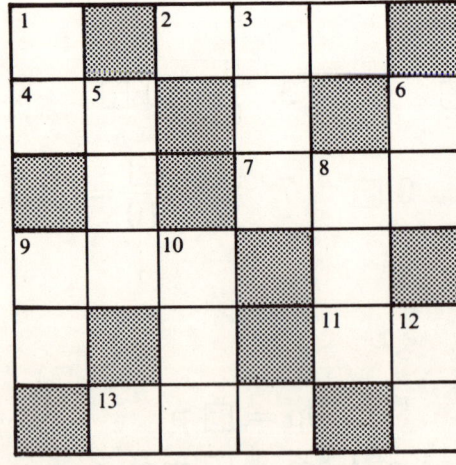

Clues down
- **1** 17 × 3
- **3** 21 × 7 × 3
- **5** 12 × 9
- **6** 20 × 4
- **8** 13 × 3 × 4
- **9** 7 × 3
- **10** 50 × 4
- **12** 7 × 7

E Find 3 numbers so that 'plus' can be changed to 'times' without altering the answers: □ + □ + □ = □ = □ × □ × □

Answers

PAGE 6

A {*1*, *3*, *5*, *7*, 9, 11, 13, *15*}

B {*2*, *4*, *6*, *8*, 10, 12, 14, *16*}

C **1.** 4 **2.** 8 **3.** 10 **4.** 14 **5.** 12 **6.** 14
7. 16 **8.** 8 **9.** 16 **10.** 12 **11.** 6 **12.** 14

D **1.** 10 **2.** 6 **3.** 16 **4.** 8 **5.** 12 **6.** 10
7. 18 **8.** 14 **9.** 14 **10.** 16 **11.** 16 **12.** 16

PAGE 7

E **1.** 7 **2.** 13 **3.** 7 **4.** 11 **5.** 15 **6.** 11
7. 15 **8.** 15 **9.** 13 **10.** 19 **11.** 19 **12.** 19
an ODD number

F **1.** 5 **2.** 9 **3.** 11 **4.** 11 **5.** 13 **6.** 13
7. 13 **8.** 15 **9.** 15 **10.** 19 **11.** 13 **12.** 17

G EVEN + EVEN = EVEN
ODD + EVEN = ODD
ODD + ODD = EVEN
EVEN + ODD = ODD

H EVEN = {*42*, 78, 80, 126, 324, 256, 728, 196, 840, 900, 374, 100}
ODD = {*53*, 61, 47, 31, 29, 259, 531, 353, 729, 333, 245, 625}
add or subtract 1

PAGE 8

A **1.** 1 hundred + 3 tens + 2 ones
2. 2 hundreds + 5 tens + 4 ones
3. 4 hundreds + 0 tens + 3 ones
4. 7 hundreds + 5 tens + 0 ones

B **1.** 2 hundreds 6 tens 4 ones
2. 5 hundreds 0 tens 2 ones
3. 6 hundreds 2 tens 5 ones
4. 7 hundreds 5 tens 4 ones
5. 5 hundreds 9 tens 0 ones
6. 6 hundreds 2 tens 1 one
7. 5 hundreds 2 tens 0 ones
8. 7 hundreds 5 tens 0 ones

C **1.** 70, 71, 72, 73, 74 **2.** 98, 99, 100, 101, 102
3. 202, 201, 200, 199, 198 **4.** 501, 500, 499, 498, 497

PAGE 9

D **1.** *Alan* 605, 428, 328, 304
Beryl 621, 571, 529, 350
Clare 640, 506, 475, 243
David 441, 347, 308, 291
Eric 675, 573, 562, 382
Fiona 728, 433, 399, 264
2. *1st table: Beryl 621, Eric 562, Clare 475, Fiona 304, Alan 291, David 291*
2nd table: Eric 675, Beryl 529, Alan 428, David 347, Fiona 264, Clare 243
3rd table: Alan 605, Eric 573, Clare 506, David 441, Fiona 433, Beryl 350
4th table: Fiona 728, Clare 640, Beryl 571, Eric 382, Alan 328, David 308
3. Fiona at table 4
4. Clare at table 2

E **1.** Z → 18 Y → 32 X → 39 W → 44 V → 47
U → 62 T → 68 S → 76 R → 81 Q → 89
2. (a) 20 (b) 30 (c) 40 (d) 40 (e) 50
(f) 60 (g) 70 (h) 80 (i) 90 (j) 100
(k) 100 (l) 110

PAGE 10

A **1.** 9 **2.** 12 **3.** 13 **4.** 11 **5.** 12 **6.** 12
7. 19 **8.** 18 **9.** 21 **10.** 21 **11.** 20 **12.** 20
13. 22 **14.** 22 **15.** 21 **16.** 21 **17.** 22
18. 26 **19.** 18 **20.** 15 **21.** 24 **22.** 26
23. 23 **24.** 27 **25.** 25

PAGE 10 (continued)

B **1.** 12 **2.** 22 **3.** 14 **4.** 24 **5.** 14 **6.** 24
7. 18 **8.** 28 **9.** 12 **10.** 22 **11.** 8 **12.** 18
13. 10 **14.** 20 **15.** 16

PAGE 11

C **1.** *4*, 9, *14*, 19, 24 **2.** *2*, 8, 14, 20, 26, 32
3. *7*, *10*, 13, 16, 19 **4.** *2*, 11, 20, 29, 38
5. *4*, 11, 18, 25, 32 **6.** *5*, 13, 21, 29, 37

D **1.** 10 **2.** 11 **3.** 16 **4.** 16 **5.** 9 **6.** 16
7. 16 **8.** 15 **9.** 15 **10.** 12 **11.** 21 **12.** 24
13. 22 **14.** 27 **15.** 15 **16.** 14 **17.** 14
18. 20 **19.** 21 **20.** 22 **21.** 15

E Arsenal *22*
Ipswich 21
Liverpool 28
Manchester U 26
Leeds 27

PAGE 12

A **1.** 5 **2.** 6 **3.** 10 **4.** 9 **5.** 10 **6.** 9 **7.** 8
8. 7 **9.** 7 **10.** 8 **11.** 13 **12.** 10 **13.** 9
14. 8 **15.** 19 **16.** 20 **17.** 15 **18.** 19
19. 16 **20.** 20

B **1.** 4 **2.** 4 **3.** 7 **4.** 10 **5.** 9 **6.** 6 **7.** 8
8. 7 **9.** 14 **10.** 6 **11.** 14 **12.** 13

C **1.** 2 **2.** 5 **3.** 7 **4.** 11 **5.** 16 **6.** 5

PAGE 13

D **1.** 5 **2.** 2 **3.** 4 **4.** 5 **5.** 4 **6.** 7 **7.** 6
8. 5 **9.** 16 **10.** 12 **11.** 17 **12.** 10 **13.** 16
14. 15 **15.** 11 **16.** 15 **17.** 19 **18.** 16
19. 9 **20.** 9 **21.** 18 **22.** 18 **23.** 11
24. 17

E Bolton *12*
Norwich 9
Spurs 6
Everton 14
Wolves 0
C Palace 14
C Palace was doing worst.

F 1.
```
6
9
10
17
19
15
```
2.
```
2
0
7
9
13
17
```
3.
```
2
7
5
1
12
13
```
4.
```
1
3
6
10
0
8
```

PAGE 14

A **1.** *4* **2.** 6 **3.** 13 **4.** 10 **5.** 9 **6.** 12
7. 11 **8.** 10 **9.** 5 **10.** 4 **11.** 8 **12.** 6
13. 8 **14.** 7 **15.** 7 **16.** 8 **17.** 4

B 1. 2. 3.

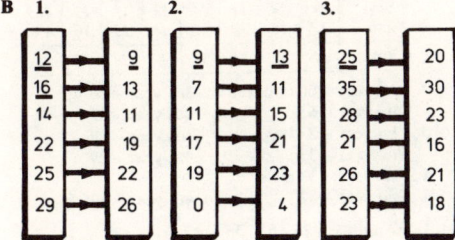

C **1.** 25 **2.** 10 **3.** 21 **4.** 14 **5.** 19 **6.** 9
7. 23

PAGE 15

D **1.** 12 **2.** 12 **3.** 16 **4.** 16 **5.** 19 **6.** 19
7. 16 **8.** 16 **9.** 22 **10.** 22 **11.** 29 **12.** 29
13. 24 **14.** 24 **15.** 28 **16.** 28 **17.** 25
18. 25 **19.** 27 **20.** 27
Yes, the same answer for each pair.

E **1.** 11 **2.** 11 **3.** 10 **4.** 10 **5.** 7 **6.** 7
7. 10 **8.** 4 **9.** 15 **10.** 3 **11.** 3 **12.** 21
13. 11 **14.** 19 **15.** 8 **16.** 14
No, each pair was not the same.

F

PAGE 16

A *t = 7 cm* x = 10 cm
u = 5 cm y = 4 cm
v = 8 cm z = 9 cm
w = 6 cm
1. x is the longest
2. y is the shortest

B **1.** 12 cm **2.** 15 cm **3.** 13 cm **4.** 17 cm
5. 11 cm **6.** 16 cm **7.** 13 cm **8.** 11 cm
9. 15 cm **10.** 9 cm **11.** 14 cm **12.** 14 cm
13. 17 cm **14.** 16 cm **15.** 14 cm **16.** 19 cm
17. 13 cm **18.** 23 cm **19.** 20 cm **20.** 18 cm

C **1.** 14 cm **2.** 14 cm **3.** 12 cm **4.** 17 cm
5. 15 cm **6.** 21 cm **7.** 22 cm **8.** 24 cm
9. 23 cm **10.** 21 cm **11.** 26 cm **12.** 28 cm
13. 24 cm **14.** 20 cm **15.** 29 cm

PAGE 17

D *h = 12 cm* j = 10 cm k = 14 cm m = 10 cm
n = 9 cm

E **1.** 2 cm **2.** 2 cm **3.** 3 cm **4.** 2 cm **5.** 1 cm
6. 1 cm **7.** 4 cm **8.** 4 cm

F **1.** 2 cm **2.** 5 cm **3.** 5 cm **4.** 1 cm

G **1.** 7 cm **2.** 3 cm **3.** 5 cm **4.** 4 cm **5.** 7 cm
6. 16 cm **7.** 9 cm **8.** 16 cm

PAGE 18

A **1.** 12 cm **2.** 15 cm **3.** 13 cm

B **1.** 13 cm **2.** 11 cm **3.** 11 cm
Isosceles triangles have two sides equal to each other.

C **1.** 9 cm **2.** 6 cm **3.** 12 cm
Equilateral triangles have all three sides equal to each other.

PAGE 19

D **1.** 9 cm **2.** 11 cm **3.** 14 cm

E **1.** 12 cm **2.** 16 cm **3.** 16 cm
A rectangle has both pairs of opposite sides equal and it has square corners.

F **1.** 12 cm **2.** 8 cm
A square has all four sides equal and it has square corners.

G **1.** 4 sides **2.** 3 sides

ANSWERS

PAGE 20

A 1. 17p 2. 8p 3. 12p 4. 22p 5. 35p
6. 29p

B 1. 5p + 2p
2. 5p + 2p + 2p
3. 2p + 2p
4. 10p + 1p
5. 10p + 2p + 1p
6. 10p + 2p
7. 10p + 5p
8. 10p + 2p + 2p
9. 10p + 5p + 2p + 1p
10. 10p + 5p + 2p + 2p
11. 20p + 1p
12. 20p + 2p + 1p
13. 10p + 5p + 2p
14. 20p + 5p + 2p + 1p
15. 20p + 5p + 1p
(or other suitable combinations)

C 1. (a) 5 2p coins (b) 2 5p coins
2. (a) 10 2p coins (b) 4 5p coins
    (c) 2 10p coins
3. (a) 15 2p coins (b) 6 5p coins
    (c) 3 10p coins

PAGE 21

D 1. 13p 2. 25p 3. 9p 4. 14p 5. 13p
6. 26p

E 1. 13p 2. 17p 3. 14p 4. 15p 5. 15p
6. 19p 7. 20p 8. 21p 9. 21p 10. 24p
11. 12p 12. 16p 13. 23p 14. 22p
15. 24p 16. 12p 17. 18p 18. 20p

F 1. 4p 2. 1p 3. 5p 4. 3p 5. 8p 6. 12p
7. 15p 8. 11p 9. 14p 10. 18p 11. 10p
12. 17p

G 1. 2p + 2p 2. 1p 3. 5p 4. 2p + 1p
5. 5p + 2p + 1p 6. 10p + 2p 7. 10p + 5p
8. 10p + 1p 9. 10p + 2p + 2p
10. 10p + 5p + 2p + 1p 11. 10p
12. 10p + 5p + 2p

PAGE 22

A 1. 10.05 a.m. 2. 10.05 p.m. 3. 6.10 p.m.
4. 4.25 p.m. 5. 6.15 a.m. 6. 2.30 p.m.
7. 8.35 a.m. 8. 11.55 p.m. 9. 7.50 p.m.
10. 8.45 a.m. 11. 10.40 a.m. 12. 5.30 p.m.

B 1. 3.05 2. 5.10 3. 8.50 4. 9.30 5. 11.45

C (a)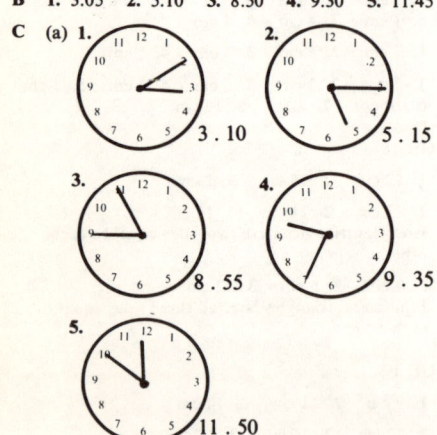

1. 3.10 2. 5.15
3. 8.55 4. 9.35
5. 11.50

(b) 1. 2.55 2. 5.00

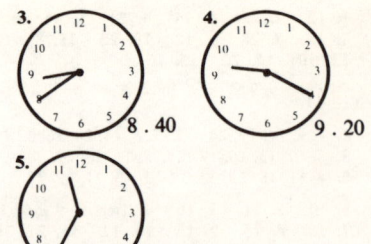

3. 8.40 4. 9.20
5. 11.35

PAGE 23

D 1. 12.05 p.m. 2. 12.15 p.m. 3. 12.20 p.m.
4. 12.30 p.m. 5. 12.35 p.m. 6. 12.40 p.m.
7. 12.45 p.m. 8. 12.50 p.m. 9. 12.55 p.m.
10. 1.00 p.m.

E 1. 5 min 2. 10 min 3. 7 min 4. 10 min
5. 5 min 6. 5 min 7. 10 min 8. 15 min

F 1. 6 min 2. 5 min 3. 7 min 4. 2 min
5. 9 min 6. 4 min 7. 33 min

PAGE 24

A 1. 58 2. 73 3. 84 4. 41 5. 31 6. 48
7. 91 8. 97 9. 65 10. 83 11. 96 12. 93
13. 52 14. 46 15. 63

B 1. 70 2. 90 3. 80 4. 70 5. 90 6. 80
7. 70 8. 100 9. 100 10. 100 11. 90
12. 60 13. 90

PAGE 25

C 1. 30 2. 70 3. 60 4. 61 5. 72 6. 47
7. 94 8. 91 9. 90 10. 90 11. 82 12. 52
13. 75 14. 98 15. 99 16. 40 17. 51
18. 64 19. 85

D France   86
Italy   85
Germany   92
Britain   91
Belgium   71
Holland   78

PAGE 26

A 1. 61 2. 22 3. 22 4. 53 5. 52 6. 73
7. 33 8. 32 9. 3 10. 7

B 1. 47 2. 12 3. 4 4. 26 5. 19 6. 63
7. 17 8. 14 9. 18 10. 11 11. 48 12. 28
13. 72 14. 36 15. 36 16. 48 17. 13
18. 17 19. 13

PAGE 27

C 1.

CLASS	1st year		2nd year		3rd year		4th year	
CLASS	1	2	3	4	5	6	7	8
boys	14	13	17	16	15	13	13	16
girls	17	16	17	15	17	15	16	17
TOTAL	31	29	34	31	32	28	29	33

2. 1st year – 60 children   3rd year – 60
   2nd year – 65   4th year – 62
3. 1st year – 27 boys   3rd year – 28
   2nd year – 33   4th year – 29
4. 1st year – 33 girls   3rd year – 32
   2nd year – 32   4th year – 33

D 1. 32 points 2. 16 3. 5
4. 1 + 4 or 2 + 3 or 5 only

E 1. 35 passengers 2. 9 empty seats
3. 54 passengers

F 1. Janet won 2. 15 seconds

G 1. 28 cm 2. 18 cm 3. 11 cm

PAGE 28

A 1. 31 cm 2. 38 cm 3. 56 cm 4. 52 cm
5. 72 cm 6. 81 cm 7. 90 cm 8. 92 cm
9. 91 cm 10. 83 cm 11. 87 cm 12. 62 cm
13. 96 cm 14. 61 cm 15. 67 cm 16. 80 cm
17. 52 cm 18. 70 cm 19. 90 cm 20. 99 cm
21. 74 cm 22. 86 cm

B 1. (a) 46 cm (b) 92 cm
2. (a) 62 cm tall (b) 69 cm (c) 71 cm
    (d) 76 cm
3. (a) 14 cm high (b) 21 cm (c) 28 cm
    (d) 35 cm 4. 74 cm long

PAGE 29

C 1. 98 cm 2. 70 cm 3. 87 cm
4. no. 2 is isosceles
5. no. 3 is equilateral

D 1. 96 cm 2. 96 cm 3. 82 cm 4. 72 cm
5. 90 cm 6. 87 cm 7. no. 1 is a square
8. no. 2 is a rectangle
(no. 3 – kite   no. 4 – rhombus
no. 5 – parallelogram   no. 6 – trapezium)

PAGE 30

A 1. 15 cm 2. 13 cm 3. 6 cm 4. 8 cm
5. 14 cm 6. 16 cm 7. 6 cm 8. 27 cm
9. 25 cm 10. 26 cm 11. 18 cm 12. 37 cm
13. 28 cm 14. 46 cm 15. 60 cm 16. 54 cm
17. 36 cm 18. 26 cm 19. 16 cm 20. 38 cm
21. 72 cm

B 1. (a) 62 cm left (b) 28 cm
    (c) 62 cm altogether
2. (a) 73 cm (b) 60 cm finally

PAGE 31

C 1. 25 cm 2. 25 cm 3. 14 cm 4. 15 cm
5. 44 cm 6. 15 cm
7. nos 3 and 4 are isosceles

D 1. (a) 37 cm wide (b) 81 cm altogether
    (c) 162 cm perimeter 2. 2 cm

PAGE 32

A 1. 60p 2. 65p 3. 57p 4. 72p

B 1. 5 10p coins 2. 10 5p coins

C 1. 20p + 1p 2. 20p + 2p + 2p 3. 20p + 5p
4. 20p + 5p + 2p + 2p 5. 20p + 10p
6. 20p + 10p + 5p + 1p 7. 20p + 20p + 2p
8. 20p + 20p + 5p 9. 20p + 20p + 5p + 2p
10. 50p 11. 50p + 2p 12. 50p + 5p + 1p
13. 50p + 5p + 2p + 2p 14. 50p + 10p + 1p
15. 50p + 10p + 5p + 2p
or other suitable combinations

D 1. 25p 2. 36p 3. 70p 4. 60p 5. 58p
6. 59p 7. 90p 8. 80p

PAGE 33

E 1. 41p 2. 51p 3. 43p 4. 90p 5. 74p
6. 77p 7. 84p 8. 81p 9. 92p 10. 72p
11. 37p 12. 82p 13. 87p 14. 88p
15. 94p 16. 73p 17. 72p 18. 78p
19. 99p 20. 72p 21. 60p

F 1. Mr Jones   p   3. Mr Rogers p
                 18   12
                 25   14
                 ──   ──
                 43   26

2. Mr Brown   p   4. Mr Collins p
              11   25
              15   15
              ──   ──
              26   40

PAGE 34

A **1.** 4p **2.** 8p **3.** 8p **4.** 8p **5.** 19p **6.** 15p
7. 24p **8.** 45p **9.** 44p **10.** 29p **11.** 38p
12. 25p **13.** 27p **14.** 66p **15.** 62p
16. 16p **17.** 23p **18.** 32p **19.** 38p
20. 47p **21.** 32p

B **1.** 25p **2.** 31p **3.** 14p **4.** 20p **5.** 44p

C **1.** (a) 15p **2.** (a) 38p
    (b) 11p   (b) 33p
    (c) 8p   (c) 22p
    (d) 3p   (d) 17p
    (e) 18p   (e) 9p

PAGE 35

D **1.** 20p + 5p **2.** 20p + 5p + 2p + 2p
3. 10p + 5p + 2p **4.** 20p + 10p
5. 20p + 10p + 2p + 2p **6.** 20p + 20p + 1p
7. 20p + 20p + 5p **8.** 50p + 5p + 1p
9. 50p + 10p + 2p **10.** 50p + 10p + 5p
11. 50p + 20p **12.** 50p + 20p + 5p + 1p
13. 50p + 20p + 10p + 2p
14. 50p + 20p + 10p + 5p
15. 50p + 20p + 20p + 5p + 2p + 2p

E **1.** (a) 32p at the first shop (b) 57p altogether
    (c) 13p left **2.** 75p

F
1. p	**2.** p	**3.** p	**4.** p	**5.** p
81	37	61	44	22
18	17	22	15	18
99	29	18	59	29
	83	101		69

PAGE 36

A 2 4 6 8 10 12 14 16 18 20

B 2 4 6 8 10 12 14 16 18 20

C (Drawings)
1. *14*, 14 **2.** 8, 8 **3.** 12, 12 **4.** 16, 16
5. 6, 6 **6.** 2, 2 **7.** 18, 18 **8.** 0, 0 **9.** 20, 20
10. 24, 24 **11.** 22, 22 **12.** 4

PAGE 37

D (number line)
1. 4 **2.** 8 **3.** 10 **4.** 6 **5.** 14 **6.** 2 **7.** 16
8. 20 **9.** 0 **10.** 0 **11.** 20 **12.** 12

E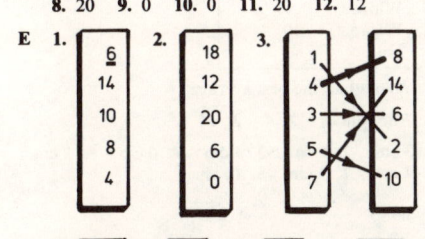

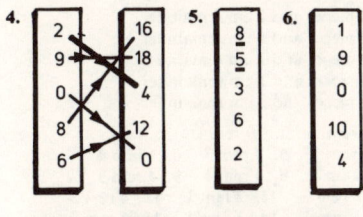

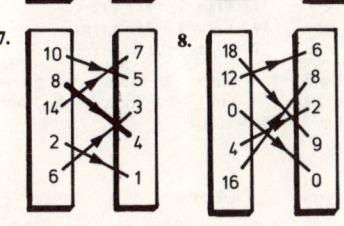

F **1.** 10 **2.** 8 **3.** 20 **4.** 12 **5.** 18 **6.** 14
7. 6 **8.** 0 **9.** 16 **10.** 4

PAGE 38

A (diagrams)

B **1.** 20, 20 **2.** 40, 40 **3.** 80, 80 **4.** 30, 30
5. 70, 70 **6.** 0, 0 **7.** 50, 50 **8.** 60, 60

C **1.**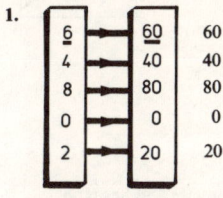

2.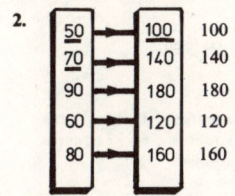

D **1.** 10, 10 **2.** 20, 20 **3.** 30, 30 **4.** 40, 40
5. 0, 0 **6.** 100, 50 **7.** 30, 15 **8.** 10, 5

PAGE 39

E *5, 10, 15, 20*, 25, 30, 35, 40, 45, *50*

F (number line)
1. $8 \times 1 = 8$ **2.** $8 \times 2 = 16$
   $4 \times 2 = 8$ $4 \times 4 = 16$
   $2 \times 4 = 8$ $2 \times 8 = 16$
3. $8 \times 3 = 24$ **4.** $8 \times 4 = 32$
   $4 \times 6 = 24$ $4 \times 8 = 32$
   $2 \times 12 = 24$ $2 \times 16 = 32$

G **1.** *2, 4, 6*, 8, 10, 12, 14, 16, 18, *20*
2. *4, 8, 12*, 16, 20, 24, 28, 32, 36, *40*
3. *8, 16, 24*, 32, 40, 48, 56, 64, 72, *80*

H **1.** 6 **2.** 12 **3.** 24 **4.** 14 **5.** 28 **6.** 56
7. 20 **8.** 40 **9.** 80 **10.** 10 **11.** 20 **12.** 40
13. 2 **14.** 4 **15.** 8

I 4, 8, 16, *double* 16 → *32*

PAGE 40

A 3 sides, 6, 9, 12, 15, 18, 21, 24, 27, 30

B 3, 6, 9, 12, 15, 18, 21, 24, 27, 30

C **1.** 3, 3 **2.** 15, 15 **3.** 18, 18 **4.** 27, 27
5. 21, 21 **6.** 30, 30 **7.** 0, 0 **8.** 9

D **1.** $2 \times 3 = 6$ **2.** $3 \times 4 = 12$ **3.** $6 \times 3 = 18$
4. $3 \times 3 = 9$ **5.** $0 \times 3 = 0$ **6.** $7 \times 3 = 21$
7. $9 \times 3 = 27$ **8.** $3 \times 5 = 15$ **9.** $3 \times 8 = 24$
10. $3 \times 1 = 3$ **11.** $3 \times 10 = 30$ **12.** $3 \times 0 = 0$

PAGE 41

E 9, 18, 27, 36, 45, 54, 63, 72, 81, 90

F **1.**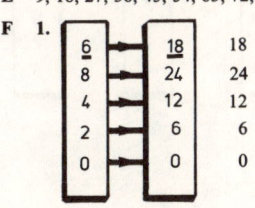

2.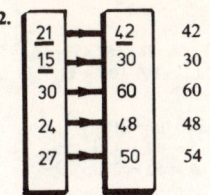

G **1.** (a) 6 **2.** (a) 9 **3.** (a) 18 **4.** (a) 27
   (b) 14   (b) 21   (b) 42   (b) 63
   (c) 16   (c) 24   (c) 48   (c) 72
   (d) 18   (d) 27   (d) 54   (d) 81
   (e) 0   (e) 0   (e) 0   (e) 0

PAGE 42

A **1.** EVEN **2.** ODD **3.** EVEN **4.** ODD
   ODD × ODD = ODD
   EVEN × ODD = EVEN

B **1.** EVEN **2.** EVEN **3.** EVEN **4.** EVEN
5. EVEN
   EVEN × EVEN = EVEN
   ODD × EVEN = EVEN

C **1.**

×	0	2	4	6	8
0	*0*	*0*	*0*	*0*	*0*
2	*0*	4	8	12	16
4	*0*	*8*	16	24	32
6	0	12	24	*36*	48
8	0	16	32	48	64

2.

×	1	3	5	7	9
1	1	3	5	7	9
3	3	9	15	*21*	27
5	5	*15*	25	35	45
7	7	21	35	*49*	63
9	9	27	45	63	81

3.

×	0	2	4	6	8
1	*0*	2	4	6	8
3	*0*	6	12	18	24
5	*0*	*10*	20	30	40
7	0	14	28	42	*56*
9	0	18	36	54	72

table 1 – EVEN NUMBERS
table 2 – ODD NUMBERS
table 3 – EVEN NUMBERS

PAGE 43

D
$$1 + 3 + 5 + 7 = 16 = 4 \times 4$$
$$1 + 3 + 5 + 7 + 9 = 25 = 5 \times 5$$
$$1 + 3 + 5 + 7 + 9 + 11 = 36 = 6 \times 6$$
$$1 + 3 + 5 + 7 + 9 + 11 + 13 = 49 = 7 \times 7$$
$$1 + 3 + 5 + 7 + 9 + 11 + 13 + 15 = 64 = 8 \times 8$$
$$1 + 3 + 5 + 7 + 9 + 11 + 13 + 15 + 17 = 81 = 9 \times 9$$
$$1 + 3 + 5 + 7 + 9 + 11 + 13 + 15 + 17 + 19 = 100 = 10 \times 10$$
$$\{1, 4, 9, 16, 25, 36, 49, 64, 81, 100\}$$

E
$$2 + 4 + 6 + 8 + 10 = 30 = 5 \times 6$$
$$2 + 4 + 6 + 8 + 10 + 12 = 42 = 6 \times 7$$
$$2 + 4 + 6 + 8 + 10 + 12 + 14 = 56 = 7 \times 8$$
$$2 + 4 + 6 + 8 + 10 + 12 + 14 + 16 = 72 = 8 \times 9$$

F **1.** $2 \times 3 = 6$ **2.** $9 \times 3 = 27$ **3.** $6 \times 3 = 18$
4. $9 \times 3 = 27$ **5.** $7 \times 5 = 35$ **6.** $6 \times 5 = 30$

ANSWERS

PAGE 44

A 1. 72 2. 77 3. 75 4. 78 5. 119 6. 95
7. 102 8. 91 9. 112 10. 104 11. 128
12. 120

B 1. *2 × 30 = 60* 2. *3 × 70 = 210*
3. *6 × 70 = 420* 4. *5 × 30 = 150*
5. *3 × 80 = 240* 6. *6 × 50 = 300*

C 1. 140 2. 240 3. 200 4. 240 5. 420
6. 560

D 1. 72 2. 100 3. 488 4. 648 5. 385
6. 423 7. 84 8. 235 9. 702 10. 696

PAGE 45

E 1.

NO. OF WEEKS	NO. OF DAYS
2	*14*
7	49
14	98
26	182
52	364

2.

NO. OF TABLES USED	NO. OF CHILDREN
2	16
7	56
9	72
12	96
14	112

3.

NO. OF ROWS PLANTED	NO. OF BULBS USED
2	32
6	96
7	112
9	144
10	160

F 1. (a) 20 cm (b) 15 cm
2. (a) 28 cm (b) 21 cm
3. (a) 36 cm (b) 27 cm
4. (a) 56 cm (b) 42 cm
5. (a) 92 cm (b) 69 cm
6. (a) 128 cm (b) 96 cm

PAGE 46

A 1. 3 sides 2. 4 sides 3. 5 sides 4. 6 sides
5. 8 sides

B 1. 2 diagonals 2. 5 diagonals 3. 9 diagonals
4. 20 diagonals
It has no diagonals.

C

SHAPE	NO. OF SIDES	NO. OF ANGLES	NO. OF DIAGONALS
triangle	3	3	0
quadrilateral	4	4	2
pentagon	5	5	5
hexagon	6	6	9
octagon	8	8	20

PAGE 47

D 1. 3 folds 2. 6 folds 3. 5 folds 4. 8 folds

E

	PERIMETERS		
LENGTH OF SIDES	regular pentagon	regular hexagon	regular octagon
7 cm	35 cm	42 cm	56 cm
12 cm	60 cm	72 cm	96 cm
18 cm	90 cm	108 cm	144 cm

PAGE 48

A (tiling patterns)

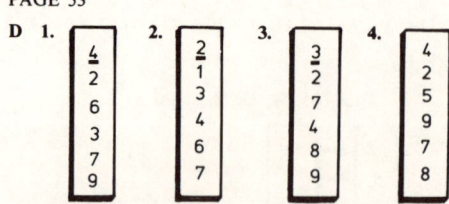

1.
2.
3.
4.
5.

B (tiling patterns)

PAGE 49

C equilateral triangle, square and hexagon
square

D no, the circle cannot be used.

PAGE 50

A 1. 8 squares 2. *5 squares* 3. 20 squares
4. 8 squares 5. 8 squares 6. 17 squares
7. 6 squares 8. 8 squares

B

SHAPE	LENGTH	WIDTH	AREA (no. of squares)
1	*4*	*3*	*12*
2	6	3	18
3	4	2	8
4	7	2	14
5	6	4	24
6	7	3	21

Area of a RECTANGLE (number of squares)
= *L*ENGTH × *W*IDTH

PAGE 51

C (practical work)

D 1. *13* unit² approximately 2. 20 unit²
3. 28 unit² 4. 20 unit² 5. 22 unit²

PAGE 52

A 1. *4 because 4 × 2 = 8* 2. *3 because 3 × 4 = 12*
3. 2 because 2 × 5 = 10 4. 2 because 2 × 6 = 12
5. 5 because 5 × 3 = 15 6. 6 because 6 × 3 = 18
7. 5 because 5 × 4 = 20 8. 3 because 3 × 8 = 24
9. 4 because 4 × 6 = 24 10. 4 because 4 × 7 = 28

B 1. 3 2. 3 3. 4 4. 9 5. 5 6. 4 7. 7
8. 2 9. 9 10. 9 11. 7 12. 8 13. 9
14. 9 15. 5 16. 6 17. 8 18. 9 19. 7
20. 4

C 1. 5 2. 7 3. 3 4. 6 5. 3 6. 4 7. 8
8. 8 9. 8 10. 9 11. 4 12. 7

PAGE 53

D 1.   2.   3.   4.

4		2		3		4
2		1		2		2
—		3		—		5
6		1		7		9
3		4		4		7
7		6		8		8
9		7		9		

E (number lines)

1. $\frac{24}{4}$ = 6 sets because 4 × 6 = 24

2. $\frac{24}{3}$ = 8 sets because 3 × 8 = 24

3. $\frac{24}{8}$ = 3 sets because 8 × 3 = 24

4. $\frac{24}{2}$ = 12 sets because 2 × 12 = 24

F 1. $\frac{36}{4}$ = 9 sets because 4 × 9 = 36

2. $\frac{36}{6}$ = 6 sets because 6 × 6 = 36

3. $\frac{36}{9}$ = 4 sets because 9 × 4 = 36

4. $\frac{36}{3}$ = 12 sets because 3 × 12 = 36

G 1. 3 sets 2. 5 sets 3. 4 sets 4. 6 sets

PAGE 54

A 1. 7 2. 4 3. 4 4. 9 5. 4 6. 3 7. 3
8. 9 9. 3 10. 5 11. 3 12. 7 13. 4
14. 9 15. 7 16. 8 17. 6 18. 2 19. 8
20. 7 21. 9 22. 9 23. 4 24. 2 25. 3

B 1. 9 2. 3 3. 6 4. 5 5. 7

PAGE 55

C 1.

BOWLER	average runs per wicket
Bowes	9
Smith	7
Barnes	6
Wicks	9
Lunn	8

Barnes had the best
bowling average.

2.

TEAM	average per game
Leeds	6
St Helen's	8
Widnes	9
Wigan	10
Hull	5

Wigan had the best average.

D 1. 7 dominoes 2. 4 3. 14

E 1. 3 cm 2. 5 cm 3. 1 cm 4. 6 cm 5. 8 cm
6. 7 cm 7. 9 cm 8. 10 cm

PAGE 56

A 1. 2 pieces and 2 cm remainder
2. 4 pieces and 1 cm remainder
3. 3 pieces and 1 cm remainder
4. 5 pieces and 1 cm remainder
5. 5 pieces and 3 cm remainder

B 1. *8 rem 1* 2. 10 rem 1 3. 4 rem 2
4. 3 rem 1 5. 3 rem 3 6. 1 rem 4
7. 2 rem 3 8. 4 rem 1 9. 4 rem 3
10. 2 rem 3 11. 2 rem 1 12. 4 rem 3
13. 3 rem 1 14. 3 rem 1 15. 5 rem 2
16. 3 rem 2 17. 3 rem 1 18. 2 rem 1

C 1. 4 rem 3 2. 4 rem 5 3. 5 rem 1 4. 5 rem 4
5. 8 rem 2 6. 6 rem 1 7. 5 rem 2 8. 7 rem 1
9. 6 rem 4 10. 9 rem 1

PAGE 57

D 1. 20 tyres 2. 3 left over

E 1.

NO. OF TYRES	NO. OF CARS	TYRES LEFT OVER
33	8	1
42	9	6
25	6	1
18	4	2
29	7	1

2.

NO. OF TYRES	NO. OF CARS	TYRES LEFT OVER
17	4	1
27	6	3
35	8	3
37	9	1
41	10	1

F

DAYS	WEEKS and DAYS	
22	3	1
30	4	2
50	7	1
60	8	4
65	9	2

PAGE 58

A 1. *1, 2, 4,* 8, 16, 32, 64, 128   2. *32, 16,* 8, 4, 2, 1

B 1. 4 cm → 8 cm   2. 7 cm → 14 cm   3. 5 cm → 10 cm   4. 2 cm → 4 cm   5. 6 cm → 12 cm

C 1. 8 cm → 4 cm   2. 6 cm → 3 cm   3. 10 cm → 5 cm   4. 14 cm → 7 cm   5. 12 cm → 6 cm

D 1. 10, 26, 34, 42, 48, 58, 72, 86, 98, 108, 118, 150   2. 10, 6, 9, 5, 3, 8, 7, 4, 11, 2, 12, 1

PAGE 59

E

SUN..	MON..	TUES..	WED..	THURS..	FRI..	SAT..
3p	6p	12p	24p	48p	96p	192p

21p by Tuesday
93p by Thursday

F

after	3 mth	6 mth	9 mth	12 mth
she had	4	8	16	32

G 1. 8 unit² (a) 4 unit² (b) 16 unit²
2. 10 unit² (a) 5 unit² (b) 20 unit²
3. 16 unit² (a) 8 unit² (b) 32 unit²
4. 10 unit² (a) 5 unit² (b) 20 unit²
5. 9 unit² (a) 4½ unit² (b) 18 unit²
6. 9 unit² (a) 4½ unit² (b) 18 unit²

PAGE 60

A 1. 19 2. 27 3. 18 4. 28 5. 22 6. 12 7. 34 8. 31 9. 37 10. 39 11. 31 12. 29 13. 36 14. 42 15. 43 16. 48 17. 49 18. 52 19. 57 20. 66 21. 36 22. 39 23. 52 24. 58 25. 54 26. 49 27. 34 28. 54 29. 58 30. 84

B 1. 13 2. 24 3. 15 4. 18 5. 17 6. 20 7. 25 8. 29 9. 26 10. 36

PAGE 61

C 1. 12 rem 4 $12\frac{4}{6}$ or $12\frac{2}{3}$
2. 15 rem 4 $15\frac{4}{6}$ or $15\frac{2}{3}$
3. 12 rem 1 $12\frac{1}{2}$
4. 18 rem 2 $18\frac{2}{4}$ or $18\frac{1}{2}$
5. 29 rem 1 $29\frac{1}{2}$
6. 24 rem 6 $24\frac{6}{9}$

7. 16 rem 4 $16\frac{4}{5}$
8. 21 rem 5 $21\frac{5}{8}$
9. 31 rem 5 $31\frac{5}{8}$
10. 24 rem 2 $24\frac{2}{6}$
11. 65 rem 4 $65\frac{4}{5}$
12. 83 rem 1 $83\frac{1}{4}$
13. 94 rem 2 $94\frac{2}{3}$
14. 31 rem 5 $31\frac{5}{7}$
15. 42 rem 2 $42\frac{2}{6}$ or $42\frac{1}{3}$
16. 36 rem 3 $36\frac{3}{6}$ or $36\frac{1}{2}$
17. 40 rem 6 $40\frac{6}{7}$
18. 51 rem 3 $51\frac{3}{6}$ or $51\frac{1}{2}$
19. 64 rem 4 $64\frac{4}{5}$
20. 71 rem 2 $71\frac{2}{6}$ or $71\frac{1}{3}$
21. 31 rem 2 $31\frac{2}{10}$ or $31\frac{1}{5}$
22. 46 rem 4 $46\frac{4}{10}$ or $46\frac{2}{5}$
23. 25 rem 8 $25\frac{8}{10}$ or $25\frac{4}{5}$
24. 64 rem 1 $64\frac{1}{10}$

The numbers have moved one place to the right.
The unit figure has become the remainder.

D 1. 41 children
2. 369 children fill the coaches
3. 7 children do not go

PAGE 62

A

NUMBER OF ROUNDS	FRACTION OF FIGHT OVER
2	$\frac{1}{2}$
4	$\frac{1}{4}$
3	$\frac{1}{3}$
4	$\frac{3}{4}$
3	$\frac{2}{3}$
5	$\frac{1}{5}$
5	$\frac{2}{5}$
8	$\frac{1}{8}$

B 1. $\frac{1}{2}$ 2. $\frac{1}{4}$ 3. $\frac{1}{3}$ 4. $\frac{3}{4}$ 5. $\frac{1}{8}$ 6. $\frac{2}{3}$ 7. $\frac{3}{8}$ 8. $\frac{1}{10}$ 9. $\frac{7}{10}$ 10. $\frac{1}{6}$

C 1. half ($\frac{1}{2}$) of

1	2	4	8	16

2. third ($\frac{1}{3}$) of

1	2	3	4	5

3. quarter ($\frac{1}{4}$) of

1	2	4	5	7

4. tenth ($\frac{1}{10}$) of

1	3	5	7	10

5. eighth ($\frac{1}{8}$) of

1	2	4	6	8

6. fifth ($\frac{1}{5}$) of

1	4	5	7	8

PAGE 63

D 1. 10 1p coins (a) $\frac{1}{10}$ (b) $\frac{3}{10}$ (c) $\frac{7}{10}$ (d) $\frac{9}{10}$
2. 4 5p coins (a) $\frac{1}{4}$ (b) $\frac{1}{2}$ (c) $\frac{3}{4}$
3. 5 10p coins (a) $\frac{1}{5}$ (b) $\frac{2}{5}$ (c) $\frac{3}{5}$ (d) $\frac{4}{5}$
4. 10 5p coins (a) $\frac{1}{10}$ (b) $\frac{3}{10}$ (c) $\frac{1}{2}$ (d) $\frac{9}{10}$

E 1. 10 mm
2. (a) $\frac{1}{10}$ (b) $\frac{1}{5}$ (c) $\frac{3}{10}$ (d) $\frac{1}{2}$ (e) $\frac{7}{10}$ (f) $\frac{9}{10}$

F 1. $\frac{6}{24} \to \frac{1}{4}$ 2. $\frac{3}{24} \to \frac{1}{8}$ 3. $\frac{4}{24} \to \frac{1}{6}$ 4. $\frac{12}{24} \to \frac{1}{2}$ 5. $\frac{5}{24}$ 6. $\frac{2}{24} \to \frac{1}{12}$ 7. $\frac{8}{24} \to \frac{1}{3}$

PAGE 64

A 1. 5p 2. 3p 3. 8p 4. 10p 5. 8p 6. 2p 7. 5p 8. 5p 9. 1p 10. 4p 11. 6p 12. 8p

B 1. 4p; 8p 2. 5p; 10p 3. 2p; 6p 4. 10p; 30p 5. 5p; 15p 6. 4p; 20p

C 1. $\frac{1}{2}$ 2. $\frac{1}{4}$ 3. $\frac{3}{4}$ 4. $\frac{1}{3}$ 5. $\frac{2}{3}$

D 1. $\frac{1}{2}$ 2. $\frac{1}{4}$ 3. $\frac{3}{4}$ 4. $\frac{1}{3}$ 5. $\frac{2}{3}$ 6. $\frac{1}{6}$

E (1. practical work)
2. (a) 6 cm (b) 3 cm (c) 4 cm (d) 2 cm (e) 8 cm (f) 10 cm

PAGE 65

F 1. (a) 6 (b) 8 (c) 6 (d) 9 (e) 12
2. (a) $\frac{3}{6} = \frac{1}{2}$ (b) $\frac{2}{8} = \frac{1}{4}$ (c) $\frac{2}{6} = \frac{1}{3}$ (d) $\frac{3}{9} = \frac{1}{3}$ (e) $\frac{4}{12} = \frac{1}{3}$
3. (a) $\frac{3}{6} = \frac{1}{2}$ (b) $\frac{6}{8} = \frac{3}{4}$ (c) $\frac{4}{6} = \frac{2}{3}$ (d) $\frac{6}{9} = \frac{2}{3}$ (e) $\frac{8}{12} = \frac{2}{3}$

G 1. 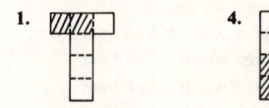   4.

2. 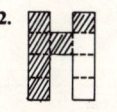   5.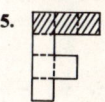

3.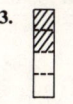

(or equivalent parts)

PAGE 66

A 2, 4, 8, 16

B 1. $\frac{2}{4}$ 2. $\frac{2}{8}$ 3. $\frac{2}{16}$ 4. $\frac{4}{8}$ 5. $\frac{4}{16}$ 6. $\frac{6}{8}$ 7. $\frac{6}{16}$ 8. $\frac{10}{16}$ 9. $\frac{14}{16}$ 10. $\frac{12}{16}$ 11. $\frac{8}{16}$ 12. $\frac{2}{2}$ 13. $\frac{4}{4}$ 14. $\frac{8}{8}$ 15. $\frac{16}{16}$

C 1. $\frac{5}{8}$ 2. $\frac{3}{8}$ 3. $\frac{3}{4}$ 4. $\frac{5}{16}$ 5. $\frac{1}{2}$ 6. $\frac{1}{4}$ 7. $\frac{13}{16}$ 8. $\frac{5}{8}$ 9. $\frac{9}{16}$ 10. $\frac{7}{16}$ 11. $\frac{3}{4}$ 12. $\frac{5}{8}$ 13. $\frac{3}{8}$ 14. $\frac{7}{8}$ 15. $\frac{3}{4}$

PAGE 67

D 1. $\frac{3}{6}$ 2. $\frac{6}{12}$ 3. $\frac{2}{6}$ 4. $\frac{4}{12}$ 5. $\frac{3}{12}$ 6. $\frac{2}{3}$ 7. $\frac{6}{6}$ 8. $\frac{12}{12}$ 9. $\frac{4}{12}$ 10. $\frac{2}{12}$ 11. $\frac{9}{12}$ 12. $\frac{9}{12}$ 13. $\frac{8}{12}$ 14. $\frac{4}{4}$ 15. $\frac{4}{12}$ 16. $\frac{6}{12}$ 17. $\frac{5}{12}$ 18. $\frac{6}{12}$ 19. $\frac{3}{12}$ 20. $\frac{9}{12}$

E 1. $\frac{6}{8} = \frac{9}{12} = \frac{12}{16}$
2. $\frac{2}{6} = \frac{4}{12}$
3. $\frac{3}{4} = \frac{6}{8} = \frac{4}{...} = \frac{6}{12} = \frac{8}{16}$

F 1. $\frac{1}{2}$ 2. $\frac{1}{4}$ 3. $\frac{3}{4}$

G 1. $\frac{1}{2}$ 2. $\frac{1}{3}$ 3. $\frac{1}{2}$ 4. $\frac{1}{4}$ 5. $\frac{1}{6}$ 6. $\frac{1}{3}$

PAGE 68

A 1. $\frac{2}{8}$ 2. $\frac{2}{6}$ 3. $\frac{5}{10}$ 4. $\frac{2}{10}$ 5. $\frac{2}{12}$ 6. $\frac{3}{6}$ 7. $\frac{4}{16}$
8. $\frac{4}{12}$ 9. $\frac{4}{8}$ 10. $\frac{2}{16}$ 11. $\frac{8}{12}$ 12. $\frac{4}{10}$ 13. $\frac{12}{16}$
14. $\frac{9}{12}$ 15. $\frac{4}{6}$ 16. $\frac{6}{8}$ 17. $\frac{6}{10}$ 18. $\frac{8}{16}$ 19. $\frac{8}{10}$
20. $\frac{6}{16}$

B 1. $\frac{1}{2}$ 2. $\frac{1}{3}$ 3. $\frac{1}{5}$ 4. $\frac{1}{4}$ 5. $\frac{1}{3}$ 6. $\frac{1}{2}$ 7. $\frac{2}{3}$
8. $\frac{3}{4}$ 9. $\frac{2}{3}$ 10. $\frac{1}{2}$ 11. $\frac{3}{4}$ 12. $\frac{3}{5}$ 13. $\frac{2}{5}$
14. $\frac{3}{8}$ 15. $\frac{1}{4}$ 16. $\frac{3}{4}$ 17. 1 18. 1 19. 1
20. 1

PAGE 69

C 1. ... $\frac{2}{8}$ and TWICE as large as $\frac{1}{8}$
2. ... $\frac{2}{4}$ and TWICE as large as $\frac{1}{4}$
3. ... $\frac{2}{16}$ and TWICE as large as ...
4. ... $\frac{4}{8}$ and FOUR TIMES as large as ...
5. ... $\frac{4}{16}$ and FOUR TIMES as large as ...
6. ... $\frac{8}{16}$ and EIGHT TIMES as large as ...
7. ... $\frac{3}{6}$ and THREE TIMES as large as ...
8. ... $\frac{3}{12}$ and THREE TIMES as ...
9. ... $\frac{2}{6}$ and TWICE as ...
10. ... $\frac{4}{4}$ and FOUR TIMES as ...

D 1. $\frac{1}{2}$ 2. $\frac{1}{3}$ 3. $\frac{1}{5}$ 4. $\frac{1}{4}$ 5. $\frac{1}{6}$ 6. $\frac{1}{10}$

E 1. ... 5 minutes → $\frac{5}{12}$ hour is 25 minutes
2. ... 30 minutes, which is $\frac{1}{2}$ an hour
3. ... 15 minutes, which is $\frac{1}{4}$ an hour
4. ... 35 minutes
5. ... 55 minutes
6. ... 45 minutes, which is $\frac{3}{4}$ an hour
7. ... 20 minutes, which is $\frac{1}{3}$ an hour
8. ... 40 minutes, which is $\frac{2}{3}$ an hour
9. ... 50 minutes, which is $\frac{5}{6}$ an hour
10. ... 10 minutes, which is $\frac{1}{6}$ an hour

PAGE 70

A

	cm	mm	cm (fraction)	cm (decimal)	mm
A → B	0	6	$\frac{6}{10}$	0·6	6
A → C	1	1	$1\frac{1}{10}$	1·1	11
A → D	2	0	2	2·0	20
A → E	2	7	$2\frac{7}{10}$	2·7	27
A → F	3	9	$3\frac{9}{10}$	3·9	39
A → G	5	1	$5\frac{1}{10}$	5·1	51
A → H	6	3	$6\frac{3}{10}$	6·3	63
A → I	7	4	$7\frac{4}{10}$	7·4	74
A → J	8	5	$8\frac{5}{10}$	8·5	85
A → K	9	9	$9\frac{9}{10}$	9·9	99
A → L	10	2	$10\frac{2}{10}$	10·2	102

B 1. 4 mm 2. 7 mm 3. 5 mm 4. 8 mm
5. 3 mm

C 1. 0·6 cm 2. 0·5 cm 3. 0·9 cm 4. 0·8 cm
5. 0·4 cm

PAGE 71

D 1. 21536·2 km or 21536 $\frac{2}{10}$ km
2. 11024·7 km or 11024 $\frac{7}{10}$ km
3. 9126·5 km or 9126 $\frac{5}{10}$ km
4. 58002·9 km or 58002 $\frac{9}{10}$ km
5. 33057·6 km or 33057 $\frac{6}{10}$ km
6. 18000·4 km or 18000 $\frac{4}{10}$ km
7. 324·6 km or 324 $\frac{6}{10}$ km
8. 18·0 km or 18 km
no. 8 was probably the newest.
no. 4 was probably the oldest.

E 1. | 5 | 2 | 0 | 4 | 1 | 7 | 10. | 3 | 2 | 4 | 4 | 5 | 6 |
2. | 8 | 4 | 0 | 2 | 5 | 3 | 11. | 1 | 5 | 6 | 0 | 8 | 4 |
3. | 0 | 2 | 4 | 0 | 9 | 8 | 12. | 0 | 2 | 5 | 6 | 0 | 8 |
4. | 1 | 0 | 0 | 0 | 0 | 0 | 13. | 0 | 0 | 4 | 2 | 8 | 3 |
5. | 0 | 5 | 1 | 0 | 9 | 2 | 14. | 2 | 4 | 2 | 6 | 4 | 5 |
6. | 0 | 0 | 1 | 1 | 7 | 1 | 15. | 6 | 0 | 6 | 0 | 6 | 9 |
7. | 0 | 0 | 0 | 2 | 5 | 5 | 16. | 1 | 2 | 5 | 4 | 7 | 1 |
8. | 2 | 5 | 0 | 1 | 0 | 9 | 17. | 0 | 6 | 0 | 0 | 0 | 2 |
9. | 6 | 7 | 2 | 4 | 0 | 2 | 18. | 1 | 7 | 5 | 2 | 4 | 7 |

F 1. 17503·6 km
→ 17503 $\frac{6}{10}$ km
→ 17503 $\frac{3}{5}$ km
2. 6214·5 km
→ 6214 $\frac{5}{10}$ km
→ 6214 $\frac{1}{2}$ km
3. 305·4 km
→ 305 $\frac{4}{10}$ km
→ 305 $\frac{2}{5}$ km
4. 6129·8 km
→ 6129 $\frac{8}{10}$ km
→ 6129 $\frac{4}{5}$ km

PAGE 72

A 1. 31·7 2. 142·1 3. 340·2 4. 273·8
5. 204·3 6. 412·4 7. 222·2 8. 52·6
9. 300·3 10. 50·3

B 1.
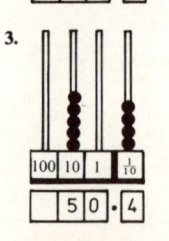
14·2

2.
114·6

3.
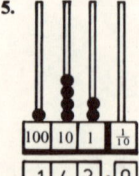
50·4

4.
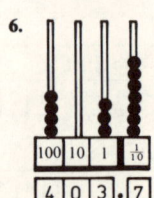
7·6

5.
142·0

6.
403·7

7.
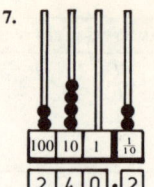
240·2

8.
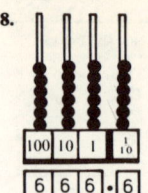
666·6

C 1.
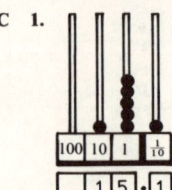
15·1

2.
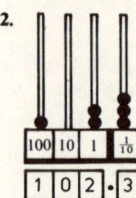
102·3

3.
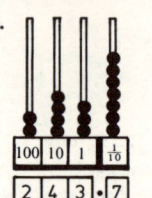
243·7

4.
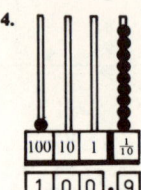
100·9

D 1.
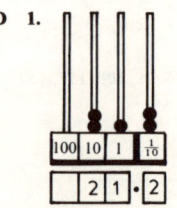
21·2

2.
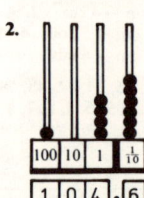
104·6

3.
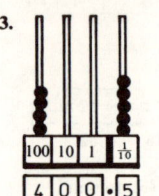
400·5

4.
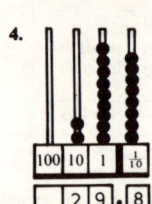
29·8

PAGE 73

E
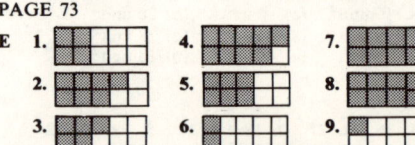

1. 2. 3. 4. 5. 6. 7. 8. 9.

	NUMBER OF SMALL SQUARES	NUMBER OF $\frac{1}{10}$'s	FRACTION IN LOWEST FORM	DECIMAL FRACTION
1.	4	$\frac{4}{10}$	$\frac{2}{5}$	0·4
2.	7	$\frac{7}{10}$	$\frac{7}{10}$	0·7
3.	5	$\frac{5}{10}$	$\frac{1}{2}$	0·5
4.	9	$\frac{9}{10}$	$\frac{9}{10}$	0·9
5.	6	$\frac{6}{10}$	$\frac{3}{5}$	0·6
6.	2	$\frac{2}{10}$	$\frac{1}{5}$	0·2
7.	8	$\frac{8}{10}$	$\frac{4}{5}$	0·8
8.	10	$\frac{10}{10}$	1	1·0
9.	1	$\frac{1}{10}$	$\frac{1}{10}$	0·1

F **1.** 3p **2.** 7p **3.** 5p **4.** 25p **5.** 35p **6.** 40p
7. 1p **8.** 1p

PAGE 74

A **1.** {*January*, March, May, July, August, October, *December*} **2.** {*April*, June, September, *November*} **3.** 217 days **4.** 120 days

B **1.** 1984, 1988, 1992, 1996
2. 1976, 1972, 1968, 1964
3. 1976, 1980, 1984, 1988 **4.** Yes

C **1.** 92 days **2.** 91 days **3.** 92 days **4.** 90 days
5. 92 days **6.** 90 days

D 45 days

PAGE 75

E **1.** *Monday* **2.** Sunday **3.** Tuesday
4. Thursday **5.** Sunday **6.** Tuesday
7. Sunday **8.** Thursday **9.** Sunday
10. Wednesday **11.** Thursday **12.** Sunday
13. Thursday **14.** Friday **15.** Sunday
16. Wednesday **17.** Sunday **18.** Saturday
19. Thursday **20.** Tuesday **21.** Sunday

F **1.** 8 days **2.** 15 days **3.** 23 days **4.** 35 days
5. 43 days

G **1.** Tuesday **2.** Saturday **3.** Thursday
4. Sunday **5.** Saturday

H **1.** 14th February **2.** 10th February
3. 7th March **4.** 28th April
5. 2nd September **6.** 11th December

PAGE 76

A **1.** (individual answers) **2.** (practical work)

B **1.** (individual answers)
2.

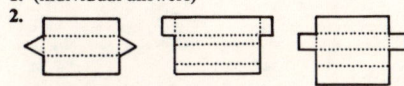

3. triangular prism → 5 faces
cuboid → 6
square end cuboid → 6
4. triangular prism → 6 vertices
cuboid → 8
square end cuboid → 8
5. triangular prism → 9 edges
cuboid → 12
square end cuboid → 12

PAGE 77

C (practical work)
prisms → constant cross section
pyramids → come to a point

D

SHAPES	RECTANGLES	SQUARES	TRIANGLES
cube	–	6	–
cuboid	6	–	–
square ended cuboid	4	2	–
triangular prism	3	–	2
square based pyramid	–	1	4
triangular based pyramid	–	–	4

PAGE 78

A **1.** 61 **2.** 65 **3.** 125 **4.** 147 **5.** 157 **6.** 24
7. 14 **8.** 44 **9.** 49 **10.** 33 **11.** 84
12. 292 **13.** 392 **14.** 584 **15.** 333 **16.** 7
17. 4 **18.** 6 rem 1 **19.** 4 rem 2 **20.** 7 rem 2

B **1.** 26 cm **2.** 36 cm **3.** 40 cm **4.** 33 cm

C **1.** 24, 6 **2.** 36, 9 **3.** 64, 16 **4.** 84, 21
5. 132, 33 **6.** 148, 37

D **1.** nearer to 80
2. nearer to 40
3. nearer to 30
4. nearer to 60
5. nearer to 60
6. nearer to 80

PAGE 79

E **1.** E → 11   **2.** E → 3
L → 7   L → 2
D → 11   D → 5
C → 8   C → 1
S → 8   S → 5
3. E → $\frac{3}{11}$ **4.** E → $\frac{8}{11}$
L → $\frac{7}{11}$ L → $\frac{5}{7}$
D → $\frac{5}{11}$ D → $\frac{6}{11}$
C → $\frac{1}{8}$ C → $\frac{7}{8}$
S → $\frac{5}{8}$ S → $\frac{3}{8}$

F **1.** $\frac{1}{2}$ **2.** $\frac{1}{5}$ **3.** $\frac{1}{3}$ **4.** $\frac{1}{4}$ **5.** $\frac{1}{3}$ **6.** $\frac{3}{4}$ **7.** $\frac{2}{5}$
8. $\frac{2}{3}$ **9.** $\frac{3}{4}$ **10.** $\frac{4}{5}$

G **1.** 0·1 **2.** 0·3 **3.** 0·9 **4.** 0·7 **5.** $\frac{2}{10}$ = 0·2
6. $\frac{8}{10}$ = 0·8 **7.** $\frac{5}{10}$ = 0·5 **8.** $\frac{6}{10}$ = 0·6

H **1.** 5p **2.** 2p **3.** 5p **4.** 1 mm **5.** 2 mm
6. 3 mm **7.** 5 mm **8.** 6 mm **9.** 8 mm
10. 1 mm **11.** 3 mm **12.** 9 mm

PAGE 80

A

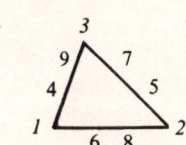

B **1.**
$$\begin{array}{r} 29 \\ +35 \\ \hline 64 \end{array}$$
2.
$$\begin{array}{r} 42 \\ -19 \\ \hline 23 \end{array}$$
3.
$$\begin{array}{r} 27 \\ \times\ 4 \\ \hline 108 \end{array}$$
4.
$$\begin{array}{r} 5 \\ 7\overline{)35} \end{array}$$

C

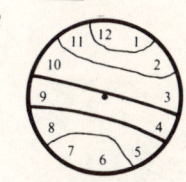

D

5	1	4	4
1	1	4	8
	0	1	0
2	8 2	5	
1	0	6	4
	1 0 0		9

E $1 + 2 + 3 = 1 \times 2 \times 3$